D1060135

REFRAMING SCOPES

Journalists, Scientists, and Lost Photographs from the Trial of the Century

reframing SCOPES

Marcel Chotkowski LaFollette

UNIVERSITY PRESS OF KANSAS

Published by the
University Press of Kansas
(Lawrence, Kansas 66045),
which was organized by the
Kansas Board of Regents
and is operated and funded
by Emporia State University,
Fort Hays State University,
Kansas State University,
Pittsburg State University,
the University of Kansas,
and Wichita State
University

Library of Congress
Cataloging-in-Publication Data
LaFollette, Marcel C. (Marcel Chotkowski)
Reframing Scopes : journalists, scientists, and lost
photographs from the trial of the century / Marcel
Chotkowski LaFollette.
p. cm.
Includes bibliographical references and index.
ISBN 978-0-7006-1568-1 (cloth : acid-free paper)
1. Evolution (Biology)—Study and teaching—Law and
legislation—Tennessee. [1. Scopes, John Thomas—Trials,
litigation, etc.] I. Title.
KF224.S3L34 2008
345.73′0288—dc22
2007047707

British Library Cataloguing-in-Publication is available.
Printed in the United States of America

10 9 8 7 6 5 4 3 2 1

The paper used in this publication is acid free and meets
the minimum requirements of the American National
Standard for Permanence of Paper for Printed Library
Materials Z39.48-1992.

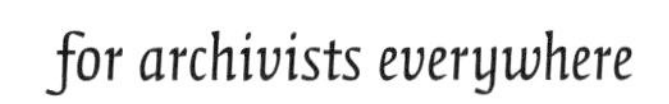

for archivists everywhere

CONTENTS

FIGURES

PREFACE

We are all actors and spectators in life. We cannot separate the two roles.
—E. Haldeman-Julius, 1928[1]

Two men in straw boaters, staring straight at the camera. The taller one, hands clasped behind his back, seems a little shy, a pleasant young man and an unlikely villain. The other, in a tweed three-piece suit, eyes glinting, hat tilted back, resembles a bantam rooster eager for a fight. With this photograph, in a jumbled pile of old negatives, lay a small brown envelope, labeled "Dayton, Tenn., June 1925—W.D.—anti-evolution trial, July 1925."

The initials and spiky handwriting on the crumbling envelope were, I recognized, those of a man whose work figured in my research for a book about science popularization. That discovery in an unindexed and partially processed collection in the Smithsonian Institution Archives launched an odyssey back through the records, first to identify the subjects of dozens of negatives and photographs and then to learn more about the people caught in the camera's lens.

The man standing next to schoolteacher John Thomas Scopes in June 1925 was George Washington Rappleyea, the instigator of a plot to bring publicity, tourists, and new business to a small town in eastern Tennessee. The photographer—science journalist Watson Davis—had traveled to Dayton to interview Rappleyea and Scopes. Davis returned in July, along with his colleague, the botanist and science writer Frank Thone, to cover the trial of young Scopes, who was charged with violating Tennessee's new law prohibiting the teaching of evolution. But, as I discovered, the story did not end there. After the trial, Davis and Thone played an important and hitherto neglected role in the life of the defendant in the "trial of the century."

Many historians have attempted to explain why *The State of Tennessee vs. John Thomas Scopes* became one of the most famous events in American life, still celebrated in popular culture and dissected in legal casebooks. Perhaps the episode seemed to epitomize the boiling over of intense religious sentiment in the South or to demonstrate contemporary social anxiety about science and modernism. Perhaps it was simply the celebrity factor—the participation of defense attorney Clarence Seward Darrow and the aging politician William Jennings Bryan. Whatever the reasons, the Scopes trial has come to serve as the consummate metaphor for cultural combat between science and religion, its meaning interpreted and reinterpreted and its lingering effects on science education fueling countless political debates.

All great events have their backstories. One of the mysteries of the Scopes trial is the defendant himself, the man whose name has become synonymous with the controversy over evolution. Even though Scopes attempted to remove himself from the public spotlight, the insistent pressure of popular culture—news articles, books, a notable play and movie—kept the mythic figure alive, sometimes casting him as villain, sometimes as hero, an enigmatic character only rarely portrayed as a normal human being. Preoccupation with protagonists Darrow and Bryan also tended to eclipse the actions of other participants. *Reframing Scopes: Journalists, Scientists, and Lost Photographs from the Trial of the Century* looks behind the familiar histories and fictionalizations and focuses on the scientists, journalists, and theologians who were entangled in the political maneuvering and ballyhoo spirit of the summer of 1925 and on those who sought to befriend an ordinary young man thrust into extraordinary circumstances.

My search to understand the friends of Scopes began with the discovery of the cache of long-forgotten nitrate negatives and a few fading prints. Some of those images, such as the photographs of Scopes, Darrow, and Bryan, were easily identified, but who were the others whose faces seemed so genial, clear-eyed, and resolute? Who were the photographers, and why had they not published these evocative images? Alan Trachtenberg has noted that historians tend to use "words, narrative, and analysis" to locate understanding whereas "the photographer's solution is in the viewfinder," such as in "where to place the edge of the picture, what to exclude, from what point of view to show the relations among the included details."[2] Fortunately, Davis and Thone used both text *and* image. Davis's scribbled notes and Thone's rough typescript drafts preserve their attempts to make sense of what they observed. Their correspondence with Scopes in the years following the trial reveals the next acts in a lifelong drama. As informal liaisons between the defense attorneys and the scientific community, the two journalists had assisted Scopes's cause at Dayton, but they also played an important role in his life afterwards. Out of sheer altruism,

FIGURE P.1. George Washington Rappleyea and John Thomas Scopes, Dayton, Tennessee, June 1925. Rappleyea (left), a native of New York, was an engineer and geologist who managed the Cumberland Coal & Iron Company. He was instrumental in suggesting that Dayton challenge the state's new antievolution statute and in persuading Scopes, the local high school athletic coach and physics teacher, to volunteer to be arrested. Buildings of the coal and iron works are visible in the background. Science journalist Watson Davis took this photograph during his visit to Dayton in June 1925 to interview Rappleyea, Scopes, and other participants in the plan to stage the "trial of the century." Courtesy of Smithsonian Institution Archives.

Davis and Thone arranged a scholarship fund so that the now unemployed schoolteacher could enroll in the University of Chicago to study geology. Moreover, they never exploited their access to him and even protected his privacy from the prying inquiries of fellow journalists.

These texts, in turn, help to illuminate the photographs, drawing attention to elements and people neglected in most histories of the trial. Davis and Thone did not just travel to Dayton and write news stories. They stayed in the same off-site quarters as the scientific witnesses and attorneys and participated enthusiastically in the defense preparations. Their candid photographs of the scientists and other visiting celebrities, of an outdoor baptism service, and of defiant liberal ministers assembled in front of a Dayton church help to ground the Scopes trial more solidly in the culture and religion of the South, to relate the events to a time and place on the cusp of change. Letters, interview notes, and drafts combine thereby with images to assist in understanding the social context of the trial and its effect on all those involved.

Here, then, is a story of shades of friendship and good will, a tale of newly minted coalitions between scientists and journalists, and a drama of actors and spectators in changing roles and changing times.

CHAPTER 1 OPENING LINES

In 1925, the nineteenth-century writings of naturalist Charles Darwin were still sparking controversy. Despite modern science's substantial contributions to industry, agriculture, health, communications, and national defense, not everyone accepted scientists' insights as inerrant or their intellectual sovereignty as absolute. Archaeology, geology, and paleontology were revealing fascinating new visions of the past. Biology, chemistry, and physics were promising future cures to age-old diseases and even imagining energy harnessed from the atom. But other scientific ideas, such as the concept of biological evolution, challenged the steadfast assurances of the Bible and contradicted the literal descriptions of life's creation given in the book of Genesis. To accept evolution and to promote it in the school curriculum were perceived, especially in the southern United States, as rejecting the status quo, questioning religion's fundamental authority, and potentially disturbing the social order.

In small towns such as Dayton, Tennessee, religion (primarily, but not exclusively, Protestant religion) played an essential role in all aspects of life. It was entrenched in social relationships, family traditions, politics, commerce, and, most assuredly, education. People met, married, and were buried at church; Sunday church suppers and holiday services governed family routines; piety and religiosity were offered to voters and customers as proof of honesty; and the school day began with prayer. Emotion, in turn, energized religion. The Chautauqua summer camps and visiting evangelists in revival tents roared out salvation at various volumes and degrees of restraint. During the warm months, undisguised expressions of belief could be observed outside most rural towns such as Dayton, where small churches and independent sects baptized the righteous in nearby streams.

Southerners were not without choices, of course, nor were they deprived of the newest technologies. Myriad cultural temptations lured them toward modernism. Radio brought jazz as well as Jesus into the living room. Daytonians could always climb into an automobile (or buggy) and drive to

Knoxville or Chattanooga to watch Hollywood movies or newsreels extolling the accomplishments of physicist Albert Einstein and other celebrities of the day.

Tradition beckoned down one road. Science—and the glittering future it promised to build—waved confidently from another. In the summer of 1925, the trial of *Tennessee vs. John Thomas Scopes* took place at that cultural crossroads.

ON THE GROUND

Evolution—the idea that living organisms developed and diversified over millions of years—had actually been taught in U.S. schools and universities for many years, but during the 1920s, there was a resurgence of efforts to restrict such instruction, primarily in southern states.[1] Legislation was introduced to prohibit the teaching of evolution in Florida's public schools and universities. Sectarian colleges in South Carolina persuaded legislators to add an amendment prohibiting the teaching of Darwinism to the general appropriations bill. Even though these legislative efforts often failed on technicalities, political pressure on state-funded universities continued to increase. By 1922, gossip was circulating within the scientific community about various "evolutionary casualties" around the country—professors of zoology and geology who had been fired for teaching about evolution.[2]

Among those leading the charge for the antievolutionists was politician William Jennings Bryan, who had become a vocal champion of the cause of Christian fundamentalism. Bryan was a "pragmatic" defender of the faith, spending "little time defending the 'truth' of the Bible and a good deal hailing its power to correct human flaws and solve social problems."[3] His opposition to evolution related both to his conviction that Darwinian-based eugenics threatened "a just society" and his concern that the teaching of evolution would erode Christian beliefs by setting up forces of chance and nature in God's place.[4] The former U.S. secretary of state also still harbored hopes of winning the presidency. Taking up the flags of religion and antievolutionism offered a convenient route to the front page.[5]

Every scientist seemed to have a different idea about how to respond to Bryan's well-publicized attacks on science. Some argued that such "outbursts" should be "treated with silent contempt."[6] Some proposed unofficial but coordinated public relations campaigns to place articles about evolution in such popular publications as the *Saturday Evening Post*.[7] Others suggested issuing statements or giving sermons about the compatibility of science and religion. The National Research Council, American Association for the Advancement of Science (AAAS), and other scientific groups were urged to, at minimum, "keep track of the legislative and ecclesiastical attacks upon the freedom of teaching in science" to help in anticipating future antiscience campaigns.[8] The AAAS had become so alarmed by nationwide developments in 1922 that it issued a special statement affirming that evolution was not "a mere guess" and declaring that legislative restrictions on teaching evolution "could not fail to injure and retard the advancement of knowledge and of human welfare."[9] Most scientific leaders agreed that "something ought to be done to meet the constant and almost uncontradicted assertions of the falsity of Evolution and the bankruptcy of Science."[10] Nevertheless, their responses remained uncoordinated and restrained.

A few prominent scientists jumped into the fight as individuals, engaging in public debates with antievolutionists or writing articles and books comparing science and religion. In spring 1923, for example, physicist Robert A. Millikan began making public speeches about religion, and he successfully persuaded dozens of prominent U.S. scientists, politicians, and theologians to sign "A Joint Statement upon the Relations of Science and Religion."[11] To men such as Millikan, reason offered the obvious response to the fundamentalists' challenge. Surely one need only declare resoundingly that science was compatible with religion, define clearly the intellectual boundaries and domains of expertise of each, and then the controversy would die away. Religion, Millikan's statement proclaimed, had the important purpose of developing human "consciences, . . . ideals, and . . . aspirations." Science's function was to develop "a knowledge of the facts, the laws, and the processes of nature."[12] Retire to your corners and then return to your separate tasks.

More politically astute scientists knew that press releases and proclamations would never dissuade antievolutionists from the fight. People care deeply about religion. They react hostilely if convinced that their beliefs are under attack. Bryan and other evangelists were skilled and forceful speakers and were perceptive readers of the public mind. They characterized "evolutionists" as the enemies of traditional Christianity and announced their determination to defeat them. Moreover, the battle was not really about who won debates in lecture halls but over who controlled the classroom.

By May 1924, legislative efforts to remove evolution from public school

curricula had grown so rampant around the country that journalist H. L. Mencken predicted sarcastically that, within the decade, "it will be a felony everywhere to mention Darwin or have his books in one's possession."[13] The antievolutionists had begun to focus attention on the consideration and approval of official biology, botany, and zoology curricula by state commissions. Science publishers were reporting pressure to remove or dampen discussion of evolution so that textbooks would "sell in all parts of the country."[14] One prominent science educator even confessed that he had subconsciously avoided use of the word *evolution* in his most recent book, although he added that "perhaps some shrewd witch hunter among the fundamentalists (lower case) may catch glimpses of the cloven hoof between the lines."[15]

IN TENNESSEE

William Jennings Bryan had made a well-publicized visit to Tennessee in 1924, and his speeches, which had begun to influence legislative efforts throughout the country, were welcomed in a state fiercely reasserting the rights of parents and taxpayers to control the public schools.[16] In Tennessee, the state government "linked patriotism and Protestantism" without apology and encouraged a certain unquestioning allegiance to a loosely defined "Old-Time Religion."[17] Public schools emphasized both Bible reading and moral instruction. By the 1920s, the schools had become the locus of intense cultural conflict throughout rural Tennessee. "To control the schools was to command the future," historian Jeannette Keith has explained.[18]

Science was only one among many perceived assaults on values, traditions, and order within the curriculum, however, and legislative efforts to remove evolution had foundered in the Tennessee legislature until early in 1925, when a Baptist farmer, John Washington Butler, drafted a statute making it "unlawful for any teacher in any of the Universities, Normals and all other public schools of the State which are supported in whole or in part by the public school funds of the State, to teach any theory that denies the story of the Divine Creation of man as taught in the Bible, and to teach instead that man has descended from a lower order of animals."[19] The bill made it through the lower house and, thanks to energetic lobbying by religious groups during a fortuitous legislative recess, was passed in the Tennessee senate and signed into law by the governor on March 21, 1925.[20]

FIGURE 1.1. William Jennings Bryan and John Washington Butler, Dayton, Tennessee, July 1925. John Washington Butler (right), a farmer who lived in the nearby Cumberland region, was in his second term as a state legislator when he submitted a bill banning the teaching of evolution in Tennessee's public schools. He is shown here with his hero William Jennings Bryan in the law offices of one of the prosecutors in the Scopes trial. Bryan's famous pith helmet is visible on the desk. Courtesy of Special Collections Library, University of Tennessee, Knoxville.

Although the Butler Act echoed the language of other proposed legislative restrictions throughout the country, the attempt to enforce it, Lawrence W. Levine has emphasized, was "not due to the actions of its friends but of its foes."[21] The newly established American Civil Liberties Union (ACLU) decided to take a stand in Tennessee, and it began advertising for a local teacher willing to test the law. Here, too, interested parties cared less about the science (or the individuals caught up in the controversy) than the general threat to academic freedom. Such legislation threatened the intellectual independence of *all* public school teachers, in all subject areas.

A DRUGSTORE CONSPIRACY

During Prohibition, the soda fountain at Robinson's Drugstore in the small eastern Tennessee town of Dayton provided a convenient gathering place for local lawyers, businessmen, bankers, and teachers. Some of these men had college educations; many were members of the region's wealthiest and oldest families. F. E. Robinson, the drugstore's owner, was the son-in-law of banker A. P. Haggard, the town's chief commissioner.[22] Robinson, whose advertisements trumpeted him as the "Hustling Druggist," was president of the school board and had the exclusive concession for school books in Rhea County.[23] He had, in fact, sold the allegedly offending textbook, *Civic Biology*, to students that academic year.[24] The drugstore group also included attorneys Sue K. Hicks and Wallace C. Haggard and Rhea County school superintendent Walter White.

Most of the drugstore regulars were native southerners. One was not. George Washington Rappleyea, manager of the Cumberland Coal & Iron Company, had been born in New York City. The amiable "Rapp" had married a Dayton native some years before and was now supervising the last gasp of what had once been a thriving mining and iron manufacturing operation. He had the soul of a salesman and a flair for attracting publicity. He proved to be a central figure in the management of the trial and a key connection to the scientific organizations that became involved in the defense.

Rappleyea is usually credited with bringing the ACLU advertisement to the group's attention, but what followed that action is obscured by myth and memory lapses. We do know that none of the "drugstore conspirators" were antagonistic to science or the teaching of evolution. Like other successful businessmen of the time, they did not reject modern conveniences or technologies. Science and engineering were to be admired.

FIGURE 1.2.
Robinson's Drugstore, Main Street, Dayton, Tennessee, June 1925. F. E. Robinson advertised himself in the local newspapers as the "Hustling Druggist." He was also president of the local school board and had the Rhea County concession to sell schoolbooks. During Prohibition, the soda fountain at F. E. Robinson Co. served as a gathering place for local businessmen, attorneys, and teachers. The plot to stage a test of Tennessee's new antievolution law was allegedly hatched there on or around May 5, 1925. Once the legal challenge was under way, Robinson and the other "conspirators" frequently posed around a soda fountain table with their volunteer defendant, John Thomas Scopes, to re-create their discussions for visiting photographers and journalists. Photograph by Watson Davis. Courtesy of Smithsonian Institution Archives.

FIGURE 1.3. Rhea County High School, Dayton, Tennessee, June 1925. The high school where John Thomas Scopes was said to have taught about evolution was less than three blocks from the courthouse where he was later tried. After the trial, supporters of William Jennings Bryan attempted to raise funds for a college honoring his memory. When a new county high school was built in the 1930s, Bryan College (now Bryan University) held classes for many years in the old school building. Photograph by Watson Davis. Courtesy of Smithsonian Institution Archives.

FIGURE 1.4.
Rhea County Courthouse, Market Street, Dayton, Tennessee, June 1925. Built in 1891, the red brick courthouse still stands today in downtown Dayton, shaded by large oak and maple trees much as it was in 1925. For the trial, the town installed loudspeakers so that visitors could sit outside on plank benches and hear the proceedings. At the front door, a wide and imposing staircase leads to the second floor courtroom, which is still in use. Scopes lived in a boardinghouse owned by the Bailey family, just two blocks away at the corner of Fourth and Market Streets. From the broad front porch, Scopes could see the place where he would be tried in July. Photograph by Watson Davis. Courtesy of Smithsonian Institution Archives.

Nevertheless, the ACLU offer presented an opportunity that coincided with Dayton's economic realities and ambitions. Why not charge a local teacher with violating this law? A public trial—and the resulting national attention—might bring thousands of visitors, invite widespread press coverage and publicity, and thereby reinvigorate the town's limp economy. It was 1920s boosterism at its best.

The regular biology teacher, also the school principal, was a married man; involvement in any controversial legal action would jeopardize his livelihood. So Rappleyea and the others persuaded twenty-four-year-old John Thomas Scopes, a well-liked physics teacher and athletic coach at the Rhea County High School, to be the defendant. As with much of life, love (or the dream of love) played a role in events. Although the school year (and Scopes's contract) had ended on May 1, Scopes had not immediately left to visit his parents in Kentucky before beginning a summer job. He had just met an attractive girl and wanted to see her again at that weekend's church supper.[25]

On May 5, Scopes volunteered to be formally arrested, and he was charged the next day. Rappleyea wired the ACLU on May 5, outlining a plan for a "four-round fight" that would eventually include a hearing before the U.S. Supreme Court.[26] A preliminary indictment was handed down May 9. As ACLU officials wrote to their newly created Committee on Academic Freedom, "The case thus originated as a friendly one, the school authorities cooperating . . . , Dr. Rappleyea . . . instigated the case solely in order that the statute might be tested and in the hope that it would be found unconstitutional."[27]

The local prosecutors immediately set about contacting William Jennings Bryan.[28] By May 12, he had agreed to come. A few days later, two of the most famous defense attorneys in the country, Clarence S. Darrow and Dudley Field Malone, made public statements volunteering their services to Scopes.

The conspirators' first action, however, had been to alert the press. News items appeared around the country within two days of Scopes's agreeing to be arrested. Soon, Rappleyea, Robinson, and the others were granting interviews and cheerfully posing at the soda fountain table with "defendant" Scopes for reporters from the biggest newspapers in the land.

CHAPTER 2 EDUCATION, PERSUASION, AND PASSION

On May 7, 1925, a brief news item appeared on the front page of the *Washington Post*: "J. T. Scopes, of the science department of the Rhea County High School, was arrested Tuesday by a deputy sheriff, charged with violating the Tennessee law prohibiting the teaching of evolution in the State public schools."[1] News of the arrest immediately attracted the attention of a young journalist, Watson Davis, and he clipped the item for his files. Here was a story that his organization, a nonprofit news agency called Science Service, would want to track.

SELLING SCIENCE

As the United States had emerged from World War I, the wealthy newspaper publisher E. W. Scripps was among many business leaders convinced that science and technology were elements essential to the nation's future growth. A healthy working democracy needed citizens with more than factory or farming skills. Easy and reliable access to scientific knowledge must be a central part of modern civic education. In 1921, in collaboration with biologist William Emerson Ritter and in cooperation with leading scientific organizations, Scripps decided to establish and endow an organization unlike anything that existed at the time—a group devoted to improving the popular communication of science.[2]

The public's needs and interests were to be paramount in all activities of this new organization, and the news content it produced had to be interesting, readable, and technically accurate. To lend authority and authenticity to its work, therefore, Science Service worked hand in hand with prominent scientists and journalists. The board of trustees included presidents and senior officers from the American Association for the Advancement of Science, National Academy of Sciences, and Smithsonian Institution and such well-known newspaper and publishing executives as William Allen White of the *Emporia (Kansas) Gazette*.

The first director, handpicked by Ritter and Scripps, epitomized a new breed of science writer. A native of Kansas, Edwin Emery Slosson had

earned a Ph.D. in chemistry at the University of Chicago while also teaching at the University of Wyoming. His literary skills soon attracted the attention of entrepreneurial publisher Hamilton Holt, who in 1903 persuaded Slosson to move to New York City and join the staff of the *Independent*. By age fifty-five, when he took over Science Service and set up its offices in Washington, D.C., Slosson considered himself a "renegade" from scientific research.[3] He was, however, one of science's foremost popularizers and promoters.

Although Scripps provided a handsome annual endowment and Science Service was set up as a not-for-profit corporation, the organization derived an essential part of its operating income from selling its syndicated articles to newspapers and magazines.[4] This mandate to "sell" science, that is, to market science news as a commodity, proved to be a crucial factor in motivating the organization's aggressive involvement in the Scopes trial, as did also the organization's pragmatic approach to news gathering, which involved exploiting local "stringers" to acquire special access to people and events.

FIGURE 2.1. Edwin Emery Slosson, director, Science Service, ca. 1925. E. E. Slosson (1865–1929) was born in Albany, Kansas, and studied chemistry at the University of Kansas (B.S., 1890; M.S., 1892) before taking a post at the University of Wyoming, where he taught until 1903. Slosson earned a Ph.D. in chemistry from the University of Chicago (1902) but became more interested in popularizing science than in research. From 1903 to 1920, he was on the editorial staff of the *Independent* in New York and in 1921 became the first director of Science Service, a nonprofit organization devoted to science popularization. An active member of the Congregational church, an evolutionist, and an ardent advocate of women's suffrage, Slosson was praised by University of Kansas dean Olin Templin as a "champion of free thought and speech for the teacher and the student." Courtesy of Smithsonian Institution Archives.

STAFFING SCIENCE

Science Service drew considerable strength and vitality in its early years from a youthful, energetic, enthusiastic, and well-educated staff. In 1925, the managing editor was twenty-nine-year-old Watson Davis, a political liberal, a modernist, and an unembarrassed booster of science and technology.[5] Son of a Washington, D.C., high school principal, Davis had grown up in a middle-class neighborhood less than a mile from the U.S. Capitol. After graduating from George Washington University with a degree in civil engineering, he married his college sweetheart, Helen Miles, and went to work testing concrete at the U.S. Bureau of Standards. Even though Davis was doing well in his government job, he harbored a strong desire to be a writer and had been moonlighting as science editor for the *Washington Herald*. When he heard that Science Service was being founded in January 1921, he immediately applied for a job.

At first, Slosson was reluctant to hire Davis full-time. The chemist believed that only *scientists* could reliably explain science to the public. Fortunately for Davis, Slosson needed help. He asked Davis to serve as part-time editor of the organization's *Science News-Letter*. Within two years, the first managing editor was fired, and Slosson replaced him with Davis.[6]

In 1925, the senior biology editor on staff was another midwesterner

FIGURE 2.2.
Watson Davis, managing editor, Science Service, ca. 1924. Watson Davis (1896–1967), the son of a high school principal, was a third-generation Washingtonian who grew up in a middle-class Capitol Hill neighborhood. Davis earned a graduate degree in civil engineering from George Washington University in 1920. Upon learning of the establishment of Science Service, he immediately applied for a job but did not become the full-time managing editor until 1923. After the death of E. E. Slosson, Davis was eventually appointed director, a post he held until 1966. Throughout his career, in addition to publishing thousands of articles and several books, Davis experimented with new communication formats and technologies, from the initiation of a popular science radio series in 1924 to early advocacy of microfilming for science documentation. He was a founding member of the National Association of Science Writers and also created the Science Clubs of America and the Science Talent Search for high school students. Courtesy of Smithsonian Institution Archives.

FIGURE 2.3. Frank Thone, senior biology editor, Science Service, ca. 1925. Frank Ernest Aloysius Thone (1891–1949) was born in Davenport, Iowa, attended Grinnell College, and earned a Ph.D. at the University of Chicago in 1922, where he specialized in plant ecology. After a few years teaching at the University of Florida, Thone decided to leave academe for a career in science writing. Thone joined the Science Service staff during the summer of 1924, and he became an active and well-liked member of the Washington, D.C., scientific community, involved in groups ranging from the Botanical Society of Washington and the Cosmos Club to the Outdoor Writers Association. As a biologist, Thone believed strongly in evolution and its importance in the science curriculum. As a devout Catholic, he believed in the compatibility of science and religion. Thone made clear connections between what he knew as a scientist, what he observed as a naturalist, and his values as a moral man, saying once that "the mark of the true conservationist is that he will respect and spare his neighbor's wildlife rights as well as his own." Courtesy of Smithsonian Institution Archives.

and graduate of the University of Chicago, thirty-four-year-old Frank Thone. A native of Iowa, Thone completed his Ph.D. in botany and worked for a while as William E. Ritter's assistant at the Scripps Institution of Oceanography. He then tried university teaching, but he soon found more satisfaction in writing popular science. In the summer of 1924, Thone left his professorship at the University of Florida and moved to Washington to work for Science Service.

Red-haired and extremely tall, a wicked wit and a dogged tennis player, Thone offered dramatic physical contrast to the rotund, prematurely balding Davis and the dignified and portly Slosson. In one aspect, however, the three men were in agreement—evolution was a force and a fact and should be taught in the schools. The previous year, Thone had led a group of Florida biologists in protesting an administrative assault on academic freedom. A devout Catholic, Thone believed passionately in the compatibility of science and religion. There is no reason, he and the other Florida professors wrote, "why scientific facts or hypotheses should undermine the beliefs of intelligent students."[7]

JUSTIFIABLE INVOLVEMENT

Scientists around the country reacted with alarm to the news that a teacher had actually been *arrested* in Tennessee for discussing evolution. Slosson immediately offered to assist the ACLU, stating in both a letter and a telegram that his organization was "anxious to keep closely in touch" with progress of the case.[8]

The antievolutionists' attacks had actually been attracting Slosson's attention for some time, but heretofore his response had been rather meek. In 1921, the dean of the University of Kansas School of Journalism had argued to Slosson that biologists were "missing their greatest opportunity to educate the American people when they allow Bryan to continue . . . to denounce them and their works," and he urged action: "Why doesn't someone who has the facts as well as the theory 'take him on' in some great newspaper. . . . Science might thus capitalize on the passing fame of Bryan. Why don't you do it yourself?"[9] In the letter's margin, Slosson had marked this latter passage with blue pencil yet had not taken up the challenge.

Slosson's hesitation is a reminder of the anxiety that permeated the scientific community during the 1920s. Religious identification and religious participation were common among middle-class scientific intelligentsia.

They did not seek open cultural warfare with religious leaders even though the censorship attempts were growing more blatant. Slosson certainly disdained efforts to pressure book publishers to "conform with the demands of the Fundamentalists in order that they might sell their books in those states," but he tended to frame the antievolution movement in the context of the failure of popular education rather than of religious hatred.[10] If the public fails to understand the character of scientific thought, then scientists must share some of the blame for failing to explain it. If science publications shy away from discussing religion, Slosson argued, then they hand victory to antievolutionists. Reasoned debate, even in the context of a legal trial, would be healthy and could contribute to civic education. Open cultural conflict was to be avoided.

In this reluctance to engage aggressively in the debate over evolution, Slosson's own religious identity undoubtedly played a role. He was active in Mt. Pleasant Congregationalist Church in Washington, D.C.; his brother-in-law, Bryant C. Preston, was a Congregationalist minister in Los Angeles; and in July 1921, Slosson was elected to the Commission on Temperance of the National Council of Congregational Churches of the United States.[11] The Emmerich Lecture Bureau, which managed the chemist's paid lecture tours, advertised one of his standard talks on "Religion and Science" as "suitable for a church forum."[12] Slosson favored rapprochement between science and religion, not cultural warfare.

When Slosson finally moved his own organization off the sidelines, he did so initially on his own terms.[13] Science Service arranged to disseminate Millikan's May 1923 press release about the "Joint Statement upon the Relations of Science and Religion," and Slosson added his own personal commentary on science's compatibility with religion.[14]

Slosson, Davis, and Thone thus each had strong personal opinions about the political and philosophical issues at stake. Nevertheless, they all recognized the impending trial as an irresistible business opportunity. Here was a chance to sell their special expertise. Let others write about politics or local color. *They* could interpret the biology, geology, and anthropology that might be addressed in the trial. *They* had special access to the best scientists to explain evolution to the public.

At first, Science Service's involvement followed a conventional route—press releases and publications such as the *Daily Science News Bulletin* for May 18, 1925, which reprinted the AAAS 1922 resolution and most recent statement on the Scopes case and listed antievolution legislation introduced since 1921.[15] Science Service also commissioned five short commentaries, "Evidences for Evolution," from such notables as geneticist C. B. Davenport and anthropologist George Grant MacCurdy and mailed them to clients on May 20.[16]

This timid first step probably reflected Slosson's own ambiguity. Many years later, after Slosson's death, Watson Davis told John Martin of *Time* that the chemist had been "honestly of the opinion that the Scopes trial was inadvisable" but had eventually decided not to "oppose" the organization's "cooperation with the defense."[17] Slosson knew that such public involvement would enhance their visibility throughout the news industry, offer increased opportunity for sales, and respond to their benefactor's mandate to publicize science widely. "Anything that you could do in the way of attracting the attention of journalists to the subject of science will naturally create a demand for your product," Scripps had consistently urged, and it will also "create a demand by editors for scientific matter generally."[18]

CLOSE ENGAGEMENT

To obtain news and original interviews with which to stimulate such demand, Science Service could not wait for scientists to reveal their insights whenever it was convenient. That might be months or even years down the road. Instead, the organization pursued potential news stories more aggressively. The Science Service writers began, on occasion, to transcend the role of neutral observer.

Both Slosson and Davis, for example, attached themselves as official "members" of various astronomy and geology expeditions while also transmitting dispatches written as if they were mere reporters on the scene. In 1924, the office had arrangements with two different anthropologists—Knud Rasmussen and Richard O. Marsh—whereby Science Service sold news stories about the expeditions to newspaper clients and (for a cut of the fee) brokered the explorers' personal narratives to popular magazines. More recently, in January 1925, Davis had joined a U.S. Naval Observatory expedition, onboard the dirigible *Los Angeles*, to observe a solar eclipse with a group of twenty-five scientists and astronomers.[19]

It was all very well intentioned. Such close engagement was perceived as assisting the organization's larger mission. Getting close to scientists and to science would make the news stories better. The notion of "objectivity," so frequently debated today, was just then emerging within journalism.[20]

Neither Slosson nor Davis nor Thone apparently saw any conflict between their roles as participants in and reporters of scientific events.

The Scopes trial promised to be a science news bonanza. Close involvement would help the organization—by enhancing its reputation and promoting its usefulness to the scientific elite—and might also benefit science. Davis began to line up clients for special dispatches from Dayton and to look for some local help.[21]

HOOKING A STRINGER

Covering the trial initially posed a scheduling dilemma for the small Science Service staff. The Rhea County grand jury session was to be held on May 25, and Judge John Raulston first announced that the full trial would take place in June. Davis had already made arrangements to leave on June 4 for a month-long train trip to the West Coast, where he intended to cover scientific conferences, visit laboratories, meet with existing newspaper clients, and solicit new business; Slosson had made major lecture commitments around the country during June, and his income from these appearances was essential to the struggling organization. Thone would have to run the Washington office. So they asked the ACLU to suggest someone in Dayton who might cover the grand jury session and trial in June for them.[22]

Locating such local correspondents (known as stringers) had become standard practice. Within his first weeks on the job, Slosson had begun to establish networks of graduate students and young scientists in cities and on university campuses all over the world. Thone, in fact, had first written for Science Service as a stringer. These part-time contributors—authorized to send telegrams postpaid with breaking news and promised modest payment for any material published—had become Science Service's extra eyes and ears.

In Dayton, they hooked onto a live one.

The ACLU suggested "Dr G W Rappleyea as your Dayton correspondent for quick news," and Davis immediately telegraphed Rappleyea, asking him to act "as our Dayton correspondent."[23] Davis instructed Rappleyea to wire accounts of any developments in the case via press collect and inquired whether Rappleyea, Scopes, and Judge John R. Neal (Scopes's local attorney) would be in town in early June, when Davis might be able to visit.[24]

Within a few hours, Rappleyea sent a ninety-eight-word telegram, primarily summarizing information already published in the newspapers and including the information that "DAYTON MEN MEET TODAY RAISE FUND OF FIVE THOUSAND DOLLARS TO TAKE CARE VISITORS DURING TRIAL PROVIDE LOUD SPEAKERS FOR COURT HOUSE SO THAT ALL PERSONS MAY HEAR EVERY WORD OF THE TRIAL."[25] Davis responded with a letter saying that "we feel that the Scopes case is important and we hope that the scientists of America will do their share in the defense of their rights" and wishing his new friend Rappleyea "heartiest good wishes . . . in your fight against Fundamentalism."[26]

SHYING AWAY FROM CONTROVERSY

Despite the staff's enthusiasm and the arrangement with Rappleyea, Science Service's direct involvement in the trial was not a foregone conclusion. Once the trial was rescheduled for mid-July, Slosson did agree that Davis and Thone should attend. But they had not yet decided how much more the organization should participate. Again, hesitation stemmed from both fear and ideology.

The trial of John Scopes promised to be far more sensational than any solar eclipse expedition. The trustees' support and endorsement would be necessary. In a memorandum describing the status of the Scopes case, Slosson and Davis asked the Executive Committee to approve expenditures of up to $500 "to pay the expenses of a competent scientist or scientists to be placed at the disposal of the defense. . . ."[27] "With Clarence Darrow and Dudley Field Malone acting as counsel for the defense," they explained, "the technically legal defense will undoubtedly be adequate, but unless further steps are taken it is likely that the scientific advice will not be adequate."

Not every trustee, it turns out, endorsed participation, even to the modest extent initially requested. One supporter, the historian Mark Sullivan, urged extreme caution, writing to Davis that he was "opposed to our cooperating with those people down in Tennessee," because such action seemed far afield from the organization's charter and might lead to "embarrassments" or "misrepresentations of our position in the newspapers" or to guilt by association. "Some of those who are going to associate themselves with that school teacher in Tennessee are good men," Sullivan added in a postscript to the letter:

> But some . . . are men not distinguished for prudence . . . all the cranks and "nuts" in the universe will become associated with it. . . . In a broad

sense perhaps it is our fight; nevertheless, the rule against purchasing an interest in somebody else's row holds good here. We are engaged in persuading people to accept science. . . . Since we are engaged in a work of education and persuasion, we run some risk in thrusting ourselves into a situation where people are involved in passion, where their minds are hardened against persuasion and education.[28]

Then, by hand, Sullivan made another (and most prescient) observation about the likely course of public debate over evolution:

> I shouldn't wonder if the newspaper headlines will represent the Modernists as (a) denying that there is a God, [and] (b) asserting that man is descended from the monkey.

Although Slosson admitted that he, too, was "apprehensive" about the ACLU's approach, he attempted to persuade Sullivan that the cause was just and their organization must go beyond merely reporting on the trial "as a piece of scientific news." The scientific community's long-standing refusal to acknowledge the threat of the antievolution campaign, Slosson argued, had allowed the fundamentalists' challenge to grow "with little opposition" and had already intimidated some publishers into eliminating discussion of evolution from their textbooks. If the ACLU proceeded as planned, that is, if the trial became a fight "purely on the question of interference with free speech," then, yes, science might suffer. A loss would reinforce the right of local and state governments to determine the content of public school curriculum and thereby leave science teachers even more vulnerable to legislated restrictions.[29] To help science avoid a "black-eye," Slosson believed, groups like Science Service must "supply ammunition" for evolution's defenders to use in public forums such as the forthcoming trial.[30]

IN MOTION

Within a week, the Science Service Executive Committee signaled its rejection of neutrality. When they met on the morning of June 2, they approved the appropriation of $1,000, twice the amount Slosson had requested, to be used "for expenses in full reporting of proceedings" and for any "problems involved in the trial."[31] The resolution stressed that "reporting" could involve "securing and presenting accurate data bearing upon the validity of the scientific evidence . . . ." The trustees did not agree to subsidize the expenses of defense witnesses (as AAAS had promised to do), but they did agree to assist the defense in presenting scientifically accurate testimony. It was a thin line.

Watson Davis left Washington's Union Station that afternoon on his planned western trip, with a brief detour to Dayton.

CHAPTER 3 DETOUR TO DAYTON

Watson Davis intended to make only a quick side trip to Tennessee that June, but he planned his excursion to "the hot bed of evolution" like a military reconnaissance mission.[1] Ever the camera buff, Davis assured Frank Thone that he would "look the ground over and see how much artillery will be necessary" to cover the trial, adding that "personally I am making the advance foray armed only with a 4.5 lens focused upon a 4.5 × 6 centimeter film pack."[2] In anticipation of his long West Coast trip, Davis had purchased a new German-made Ica Victrix camera, with a portable developing tank to allow him to develop his negatives as he traveled.[3]

Instead of hostile territory, Davis found a tranquil town and friendly natives. Even the acerbic columnist H. L. Mencken proclaimed Dayton "full of charm and even beauty," with an abundance of "cool green lawns and stately trees."[4] In the photographs Davis took during that first visit, the town appears a trifle sleepy in the warm spring air. No sightseeing buses or tourists disturbed the small business district. Later, there would be a huge banner pointing to Robinson's Drugstore ("Where It All Started"), but in June the flags draped lazily across Main Street advertised performances by a visiting Redpath Chautauqua troupe. Dayton resembled many other southern towns—a comfortable place to live if you belonged and probably a congenial place to visit even if you didn't.

Davis lodged at the Hotel Aqua, a few blocks from the train station, and made advance reservations to stay there again during the upcoming trial. He made brief excursions into the surrounding countryside, inspecting the dormant Cumberland Coal & Iron Company operations and learning about the region's geology. He photographed the drugstore, the new high school where evolution had been taught, and the county courthouse where the trial would be held, and he interviewed John Scopes, George Rappleyea, and other participants in the unfolding drama.

FIGURE 3.1.
Main Street, Dayton, Tennessee, June 1925. This close-up of the central business block along Dayton's Main Street shows (from right) the Chevrolet dealership, the Hotel Aqua, and (visible under the flags advertising a visiting Redpath Chautauqua group) the sign for Robinson's Drugstore. Watson Davis stayed in the Hotel Aqua during his June visit and planned to stay there again with the visiting press corps when he returned to cover the trial. Photograph by Watson Davis. Courtesy of Smithsonian Institution Archives.

SCOPES

"We are all actors and spectators in life," E. Haldeman-Julius gave as one of his life lessons. "We cannot separate the two roles," he wrote. "A man does not act in one spirit and look at the rest of life in another spirit; his actions and his reactions as an observer of the passing show are closely bound with one another."[5]

John Thomas Scopes exemplified that consistency throughout his life. He had just completed his first year as a high school teacher when he agreed to be arrested. His goal had been to make enough money teaching so that he could continue his education in a few years, possibly in the law, possibly in the sciences, and he had been planning to earn extra cash that summer as an automobile salesman. He was not even a Tennessee native. The son of Mary Alva Brown and Thomas Scopes, young John had been born August 3, 1900, in Paducah, Kentucky. His father, an English immigrant, was a railroad worker, and during John's childhood the family lived in Kentucky and then Illinois.[6] After graduating from the University of Kentucky in 1924, with concentration in physics and chemistry, Scopes applied for teaching jobs in the region.

Life twists and then turns. The post at Rhea County High School in Tennessee had been advertised at the last minute, when the man who had been serving as athletic coach and physics instructor received a better offer elsewhere. Scopes later claimed, with characteristic modesty, that the school board offered him the job because he had been the "first applicant with the proper qualifications," but his science background was undoubtedly attractive, because he could substitute as needed for the regular biology teacher.[7] Scopes freely admitted that he never actually taught about evolution, addressing the subject only whenever it had come up "incidentally" in the textbooks assigned by the state board.[8] As Davis and many others observed, Scopes was a "complacent member of the conspiracy."[9]

Without question, Scopes had made many friends in town that year, especially when he had coached the high school basketball and football teams to respectable records. He was consistently described as an amiable and pleasant person. Sandy-haired, with clear blue eyes, the lean and lanky Scopes had "a quick, easy way of expressing himself" and "a slow, pleasant smile."[10] One reporter described the young man's voice as a "soft, dulcet drawl," although he added that the teacher was definitely not "an orator by nature . . . and is a better listener than he is a spellbinder."[11]

He impressed many of the famous people he met that summer with both

FIGURE 3.2.
John Thomas Scopes, Dayton, Tennessee, June 1925. Scopes (1900–1970) was in his first teaching job after graduating from the University of Kentucky in 1924. He taught algebra and physics, coached the basketball and football teams, and occasionally took over biology classes for the regular instructor at Rhea County High School. Scopes volunteered to be "arrested" for teaching evolution, even though he had not taught that part of the course. After the trial, thanks to a scholarship fund raised by Watson Davis, Frank Thone, and several of the scientists who served as expert witnesses, Scopes attended graduate school at the University of Chicago. He eventually became a geologist in the oil industry, shunning publicity and celebrity most of his life. Photograph by Watson Davis. Courtesy of Smithsonian Institution Archives.

his intelligence and his unassuming demeanor. Davis, Thone, Clarence Darrow, and the visiting scientists all described Scopes as polite, contemplative, and remarkably self-composed.[12] His participation in Dayton's plan was, in fact, both considered and characteristic, for he endorsed strongly the principles at stake, something Scopes later attributed to his father's values and political convictions. As Darrow wrote, he was "a modest, studious, conscientious lad . . . brought up by his family to have [his] own opinions, and to stand by them."[13] Even hard-boiled journalists such as the *Chicago Tribune*'s Philip Kinsley expressed admiration for a man "nervous but undaunted" in the face of the upcoming trial and "passionately sincere in his conviction that the evolution law is a form of mental slavery."[14] Defense attorney Arthur Garfield Hays later said that Scopes had "handled himself with much the same decorum that was shown by Lindbergh after his flight."[15] And even one of Bryan's loyal supporters expressed admiration for the "twenty-four-year-old boy [who] has been dignified beyond all measure of his deserts through the notoriety which has been heaped upon him."[16]

What Scopes did not relish was his new status as a celebrity, and he apparently had no desire to exploit or extend it. Journalist Raymond Clapper wrote that Scopes "has been under close observation for two weeks and the verdict of this writer and most others who have reported on the trial is that he is a highly intelligent, alert young man who wants a chance to make something of himself, but who has the good taste not to go about it by commercializing his fame to sell Florida real estate or deliver chautauqua lectures."[17] Many of the contemporaneous accounts, in fact, describe how Scopes consistently scurried away from promises of fortune.[18] By late May, reporters recognized that this defendant did not follow the normal pattern; instead of seeking the camera's lens, he stood calmly in the center, willing to do his part whenever required but far more comfortable as an observer on the sidelines. Scopes had "lately become disconcerted by his fame," one Associated Press journalist wrote in late May: "His most ardently expressed wish is that he shall not be considered 'a publicity hound.' He went on record early in the discussion as having no desire to 'break into the front pages,' but, on the other hand, was filled with an abiding desire to sell automobiles, his summer calling."[19]

As the trial date approached, Scopes became more reticent in public. During May and June, however, he cheerfully shared his opinions on evolution, the court case, and academic freedom with all journalists: "I thought

I'd lose my job, which was nothing to lose, and that it would be just a little trial like any other law suit at the courthouse. . . . It looks like a fine mess I got myself into."[20] He was young and idealistic, undeterred by the threat of attack or the whirlwind of publicity: "It has been my idea all my life to make plans and keep working toward them. And then when the plans blow up and some big thing interrupts, to seize the next best thing promptly and without fear."[21]

For newly minted celebrities in the 1920s, the fusion of person and myth, of symbol and issue, occurred seemingly overnight. As Roderick Nash has explained, the press encouraged a "cult of the hero"; it promoted those who seemed to provide "living testimony of the power of courage, strength, and honor."[22] The unassuming yet daring and intelligent Scopes thus provided "fertile ground for mythmakers." It is characteristic of the man, however, that he refused to cooperate in the process. The *New York Times* told its readers that Scopes "seemed to wonder what all the excitement was about . . . he seemed to think he should be taken merely for a young teacher in a mess, rather than a central figure in a fight for liberty of expression and teaching."[23]

"RAPP"

The architect of Dayton's scheme and of Scopes's newfound celebrity, George Washington Rappleyea, offered dramatic contrast in appearance, background, motivation, and lust for fame. Six years older and far more worldly wise, the 5 foot 4 inch Rappleyea seemed diminutive and slightly exotic next to the lanky Kentuckian. One Dayton visitor described Rappleyea as "an untidy person with rather ill-tended teeth . . . in complexion olive to the point of swarthiness" and with "thick, bushy, jet black hair . . . liberally sprinkled with grey."[24] But that same observer also praised the New Yorker's intelligence. Well-read, interested in the world around him, argumentative, alert to science and trained in engineering, his eyes sparkling with amusement and vitality, Rappleyea probably seemed a familiar type to Davis, much like the artists, writers, and scientists the journalist knew in Washington and New York.

Rappleyea's boyhood had been quite different from that of Scopes or Davis, however. A descendant of French Huguenots who had come to the United States in the seventeenth century, Rappleyea had been born in the urban tangle of New York City on July 4, 1894, and told of having worked as a newsboy in Times Square. He had spent much of his youth along the

FIGURE 3.3.
John Thomas Scopes and George Washington Rappleyea, Dayton, Tennessee, June 1925. A few days after this photograph was taken, Scopes traveled by train on his first trip to New York City, where he, Rappleyea, and Tennessee attorney John R. Neal met with Clarence Darrow, American Civil Liberties Union officials, and prominent scientists concerned about the upcoming trial. Photograph by Watson Davis. Courtesy of Smithsonian Institution Archives.

East Coast, learning to sail, stunt riding in hot-air balloons, even flying a monoplane.[25]

Davis's notes refer to "Dr. G. W. Rappleyea" as a "sanitary engineer," and Rappleyea boasted to journalists that he had degrees in bacteriology and engineering from two universities, but 1918 draft records and the 1920 census list Rappleyea's occupation as "civil engineer" and the traces of extensive graduate education remain faint.[26] As evident in numerable patents for a range of innovations and inventions, Rappleyea was clever, intelligent, and clearly knowledgeable about geology, chemistry, and engineering.

Like Thone and many other young men, Rappleyea had been mustered into the military during the last stages of World War I. He was commissioned as a second lieutenant in the U.S. Army in 1917 and sent to training in Georgia, but he served only a few months stateside before the war ended. At the time of trial, Rappleyea was still a member of the Sanitary Officers' Reserve Corps and the post commander of Dayton's American Legion chapter.[27]

After the war, Rappleyea worked as a civil engineer based in Wilmington, North Carolina, near his mother's residence in Southport, and was probably in business for himself. In 1919, he met and married a Dayton native, Ova Corvin (whom everyone called "Precious"). They moved to her hometown around 1922, where Rappleyea began managing the Cumberland Coal & Iron Company and surveying the coal, gas, and mineral resources in the area.[28]

All published accounts of Rappleyea and all known correspondence throughout his life give the impression of a person who was genial, convivial, and socially adept, great fun to be around—and a thorn in the flesh of pomposity and sanctimoniousness. The *Kansas City Star* called Rappleyea "a combative little rooster, with a shock of iron-gray hair, a passion for study and a habit of getting into arguments."[29] Philip Kinsley described him as a keen-eyed man "of studious countenance and ease of expression, a forceful debater."[30] Even Mencken praised Rappleyea's "skill and culture," noting that "the visiting scientists found pleasure in his company."[31] Publisher and writer Marcet Haldeman-Julius declared Rappleyea to be "interesting" because his "mind is essentially a scientific one, clear, disciplined; his mental integrity and intrinsic sincerity obvious."[32] For the upcoming trial, Rappleyea served as arranger and the defendant's friend-in-need (if not always in deed). He was "impresario" of Dayton's show.

Unlike Scopes, Rappleyea reveled in the publicity. By June 1, he had been interviewed and photographed; his statements had appeared in major newspapers all over the country. *Time* magazine added to the notoriety by titling one of its stories "Rappleyea's Razzberry."[33]

During June, Rappleyea escorted Davis around the countryside and gave him an insider's tour of the idle blast furnace and mines as well as a lesson on the region's geology. The declining fortunes of that coal and iron business were, in fact, a critical element in the Scopes trial.

COAL, IRON, AND STRAWBERRIES

Dayton is nestled between Walden Ridge and one of the sweet loops of the Tennessee River. The ridge extends for many miles along the eastern edge of the Cumberland Plateau, and although the cliffs do not loom over the town, they add to the pleasantness of the town's natural setting.

The climate, topography, and topsoil have favored agriculture in the bottomlands along the river; other natural resources fueled Rhea County's prosperity in the mid-nineteenth century. The area, Tennessee historian W. C. Morgan has observed, possessed a fortuitous combination of "coal for making coke to smelt ore, iron ore for smelting, limestone for fluxing melted iron and building roads, timber for lumber and mine props, clay for making bricks, fire clay for firebricks, cheap land, abundant labor, and the railroad and river for transportation."[34] Early European settlers had built forges to smelt iron ore. By the 1870s, brick making and iron making operations were well established, and many American and European investors had dreams of extracting their fortunes out of Rhea County's geological riches. During the 1870s, railroad lines from Ohio and Georgia were under construction and converging in nearby Chattanooga.

In 1879, an enterprising Englishman, Titus Salt, purchased thousands of acres of land and mineral rights and by 1883 had incorporated Dayton Coal and Iron Company, Ltd., in England and begun operations in Tennessee.[35] Titus and his fellow investors opened coal mines, built enormous furnaces, and imported great train engines, investing millions of dollars in the venture.[36] They constructed housing for workers and a large mansion for the manager. Their business contributed to a tripling of local land values. Within five years, Dayton's population had grown from 250 to 5,000; hundreds of people had found employment in the mines, at the furnace and coke ovens, and in various supporting enterprises. By 1890, Dayton had become the county seat. Beginning with the first large-scale

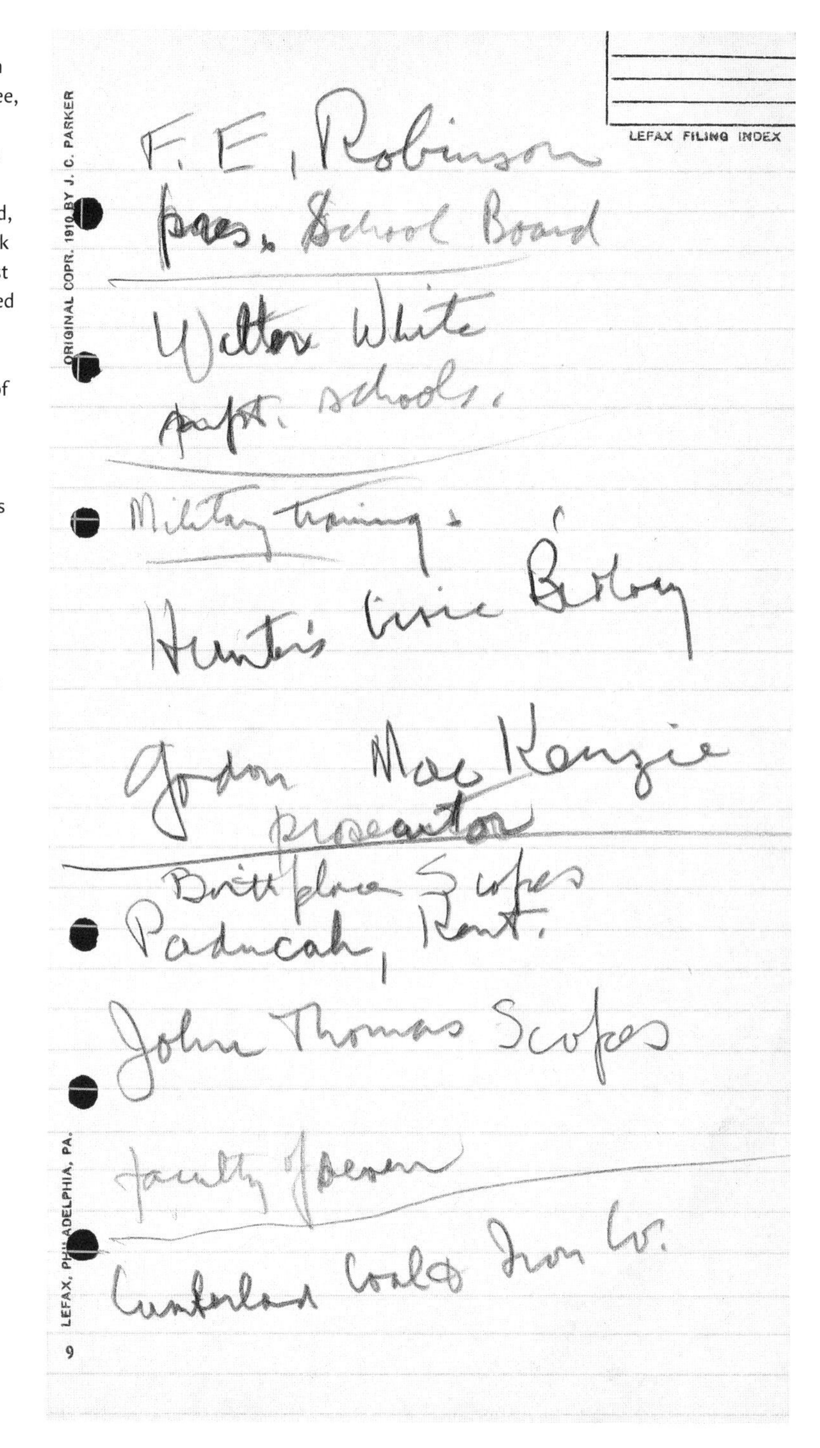

ORIGINAL COPR. 1910 BY J. C. PARKER

LEFAX FILING INDEX

F. E. Robinson
pres. School Board

Walter White
supt. schools.

Military training

Hunter's Civic Biology

Gordon MacKenzie
prosecutor

Birthplace Scopes
Paducah, Kent.

John Thomas Scopes

faculty

Cumberland Coal & Iron Co.

LEFAX, PHILADELPHIA, PA.

9

FIGURE 3.4.
Interview notes, Watson Davis, Dayton, Tennessee, June 1925. Science journalist Watson Davis routinely took notes in his Lefax, a pocket-sized, loose-leaf ring notebook used at the time by most engineers. Here he jotted down some essential information about the proposed prosecution of the schoolteacher John Thomas Scopes. Other pages from this group of notes show that Davis asked Scopes about his education and the record of the football and basketball teams that year. Courtesy of Smithsonian Institution Archives.

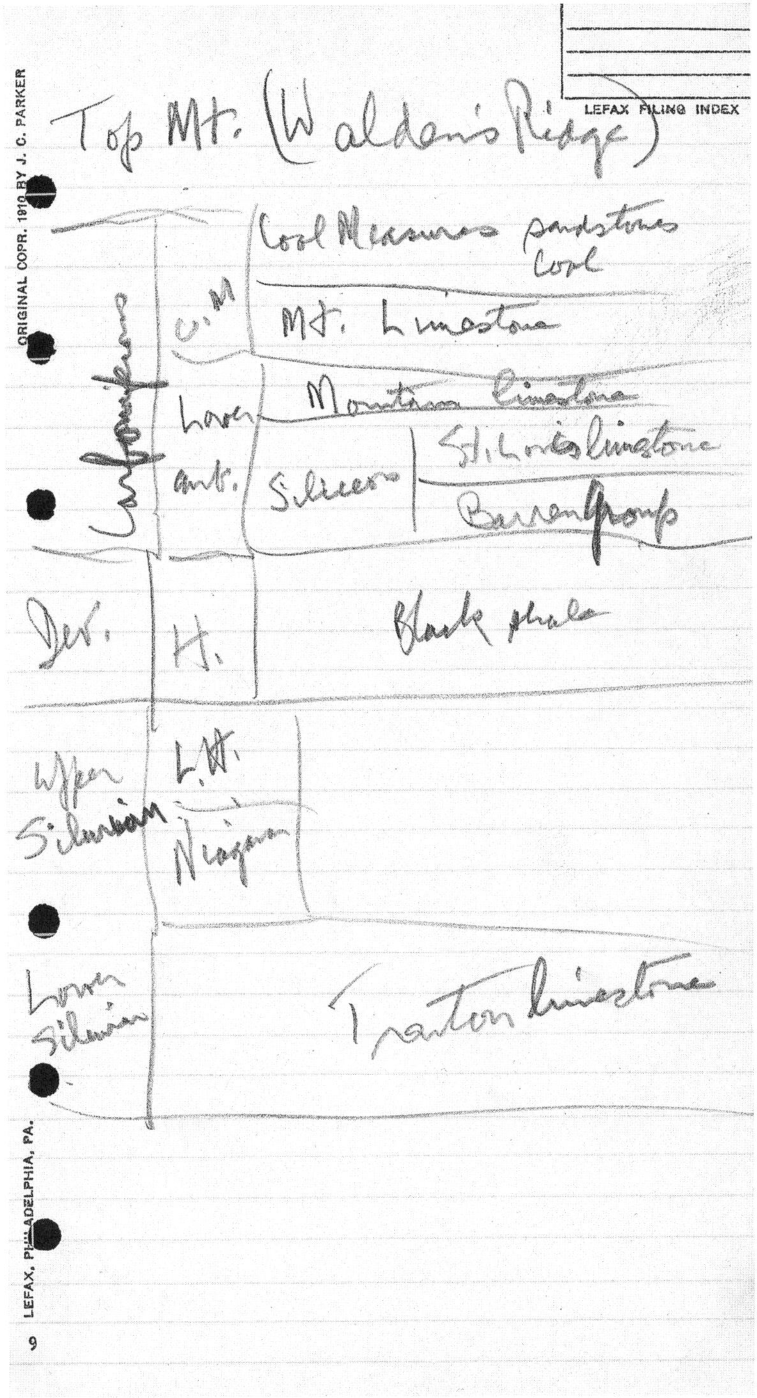

FIGURE 3.5.
Interview notes, Watson Davis, Dayton, Tennessee, June 1925. This diagram of the region around Rhea County, Tennessee, reflects some of the survey work that George Washington Rappleyea had done of the geology of Walden Ridge. Davis incorporated some of this information into the articles he wrote that June. Courtesy of Smithsonian Institution Archives.

plantings in 1898, Rhea County had also become a large agricultural producer of strawberries, shipping hundreds of railcar loads every year.[37]

Unfortunately, beginning in the 1890s, a series of financial crises and managerial upheavals, accelerated by pressures in the U.S. economy, resulted in several reorganizations and eventually the collapse of what Titus Salt had built. By 1913, the Dayton Coal and Iron Company had failed; receivership, bankruptcy, sale, and more reorganization followed. In 1921, the land, quarries, and coke ovens were sold again to a group of New York investors, who also went out of business.[38] In 1925, one more attempt was being made to restart the mines and furnace—the Cumberland Coal & Iron Company operation being managed by Rappleyea—but its prospects did not look good. Rhea County needed new sources of income. Dayton's population was hovering around 2,000. Strawberry farming, grain mills, and small textile factories alone could not support the growth that Dayton businessmen envisioned nor help them compete with other towns in eastern Tennessee.

Like many other middle-class Americans in the 1920s, the members of the Dayton Progressive Club perceived publicity as an expedient way to attract investment. Advertising men were the century's new "apostles of modernity."[39] Public relations, publicity making, and selling increasingly shaped the plans of ambitious groups and locales. Watson Davis's own organization, Science Service, was skillfully adopting the techniques of public relations to promote science to the public. For Dayton to see publicity as a route to economic development was not that unusual. The town became, as *Time* so colorfully explained, "intoxicated with 'boom' elixir."[40]

The assumption that publicity would put Dayton on the map and attract new industry inspired one of the most unusual publications associated with the trial, a paperbound booklet called *Why Dayton—Of All Places?* Published by druggist F. E. Robinson and businessman W. E. Morgan, the promotional gimmick included quotes from Sinclair Lewis's *Main Street* and cited Cincinnatus and Aristotle alongside photographs of the town's hosiery mill, feed and grain operation, and blast furnaces. The cover drawing of an arrow pointing from the clouds (from heaven?) toward the map location of Dayton was accented by a few strawberries to denote the region's agricultural prowess. The text spoke of Aristotle's anticipating the principles of evolution and of the hills of Ancient Greece, the "Cross of Calvary," and the Roman Forum as historical company for "diminutive"

FIGURE 3.6. Cumberland Coal & Iron Company buildings, Dayton, Tennessee, June 1925. The mines and factory managed by George Washington Rappleyea were the remnants of the once extensive Dayton Coal and Iron Company, Ltd., established in 1883 by English investors. Photograph by Watson Davis. Courtesy of Smithsonian Institution Archives.

Dayton. It described the drugstore conspirators as "assembled alchemists at the hearth" concocting "panaceas for human ills." And it trumpeted Dayton's bipartisan spirit and claimed that "Life in Rhea County is accomplishment through Diversity," yet failed to mention that, according to Tennessee law, "diversity" must not include evolutionists in the classroom.[41]

A noticeable tone of good-natured joshing permeated the text, as if its authors' tongues were planted firmly in their cheeks. The account of Scopes strolling into the drugstore after a tennis game and admitting spontaneously that he had used the offending textbook is even described as "half playful, half serious": "With a little sparring, a little side-stepping, the hats were in the ring, and the whole world has done its bit to hold up the stage for public view." Only the final chapter ("Dayton's Plan") revealed the trial's real purpose: "Dayton proudly rises at the banquet table of this Feast of Reason and starts off by saying: '*We are the greatest strawberry growers on earth!*' "[42] Hopes were high that a sensational event pitting fundamentalism against modernism would restore the prosperity of the past.

This scheme apparently seemed so attractive that Dayton had to fight off several attempts by nearby Knoxville and Chattanooga to capture the trial, publicity, and anticipated tourist boom.[43] On May 18, four students at Chattanooga's Central High School were summoned before a grand jury and asked if *their* instructor might have violated the antievolution statute.[44] On the following day, a Chattanooga newspaper even suggested that moving the Scopes trial from Rhea County to Hamilton County would be best because spectators could be more "comfortably" accommodated in that town's plentiful hotels and new 1,000-seat Memorial Auditorium than in the smaller Dayton venues.[45]

TESTIFYING FOR EVOLUTION

Most journalists who ventured to Tennessee that June tended to focus on the anticipated confrontation between Bryan and Darrow or on how the town was coping with unaccustomed publicity. In his accounts, Davis emphasized the science and the region's natural resources. His raw interview notes include sketches of layers of geological time, probably based on Rappleyea's own survey drawing.[46]

The town "could not be better placed geologically," Davis wrote, for "a test to determine whether natural law, made by God, or legal law, made by man, shall prevail." "The very ground" on which the courthouse sits, the underlying rocks containing "the embalmed life of ages ago," and even the

FIGURE 3.7. Rural countryside near Dayton, Tennessee, June 1925. For an amateur photographer, Davis occasionally exhibited creative flair in his composition and choice of subject. This photograph offers a bucolic view of the countryside surrounding Dayton — a rocky creek in the foreground, factory worker housing arrayed along a dirt road, and a whitewashed church visible on the hill at right. Photograph by Watson Davis. Courtesy of Smithsonian Institution Archives.

FIGURE 3.8. View of Dayton, Tennessee, June 1925. Visible in the foreground are part of the coal and iron company operations and, in the midst of a group of trees at right, the light-colored roof and brick steeple of the Rhea County Courthouse. Photograph by Watson Davis. Courtesy of Smithsonian Institution Archives.

hills surrounding Dayton all "testify for evolution." To the town's west is Walden Ridge, containing "layer upon layer of rocks each representing different and progressively older deposits." "This is the record of the rocks," Davis explained, and "close by the coal seams are layers of iron ore . . . and lower down in the hills and earlier in age are strata of limestone. . . . Below them there are Devonian and Silurian rocks and earth upon which the town of Dayton itself rests and out of which spring the strawberries and peaches, the principal products of the region."[47]

Like many on all sides of the debate, Davis could not refrain from drawing his own moral lesson from the circumstances. Rappleyea, he wrote, had learned in his work "the reliability" of this record. And Scopes, too, "knows how to read the story chronicled in limestone, shale, iron and coal"—a knowledge that "makes him more determined that he shall not be prevented from teaching his eager young pupils the most basic, most interesting and most fundamental facts of nature."[48]

SURVEYING THE COASTS

Within a few days, Davis resumed his train trip west, visiting Denver, Seattle, and San Francisco and even experiencing an earthquake in late June as he rode through Santa Barbara, California.[49] It was shaping up to be quite a year.

Scopes and Rappleyea traveled east, taking the train with Rappleyea's wife and sister-in-law to New York City to meet with scientists and defense lawyers, plot strategy with the ACLU, and confront more of the press. Reporters had already begun the process of attempting to construct Scopes as an ordinary young man thrust into extraordinary circumstances, to make his persona more familiar to readers.[50] "Johnny Scopes" could well be *your* relative or neighbor or friend. Within barely a month, he had become one of the most famous celebrities in the United States.

Despite the pressure, Scopes apparently did not let the eastern journalists rattle him. Westbrook Pegler published a colorful account of one "fast mental workout" that took place in a downtown hotel room as reporters hurled questions at the two Dayton men. Scopes "handled himself well," Pegler admitted. He "fielded the ball . . . and didn't get into any arguments."[51]

Everyone also had advice for the young man, of course. Prominent scientists such as evolutionist Henry Fairfield Osborn, president of the American Museum of Natural History, warned Scopes "not to let radicals run the case."[52] And the ACLU and many others had strong opinions about who should or should not be on the defense team. Scopes held firm and—probably influenced by his union-organizing father—chose Clarence Darrow, one of the best-known liberal lawyers in the United States.

On the way home, on June 12, 1925, Scopes and his local attorney, John R. Neal, who had joined in the New York discussions, stopped in Washington, D.C., for a brief afternoon of sightseeing before taking the night train to Tennessee. Even though the visit was quiet and unadvertised, Capitol Hill was, as usual, full of journalists, and the teacher's face had become well known.[53] On sunny June days, the windows of the upper gallery of the Library of Congress offer a spectacular view of the statue of "Freedom" atop the U.S. Capitol. While in the library, Scopes looked at an original copy of the Constitution. Then he and Neal, trailed by reporters, walked over to the Capitol and peered inside the chamber of the Supreme Court, where they imagined their case might eventually be argued.[54]

The stage was set.

CHAPTER 4 PARTICIPANTS AND OBSERVERS

The "scrap" in Tennessee may have been, as Watson Davis wrote, a cultural struggle between modernism and fundamentalism and yet another step in "the age-long fight between progress and conservatism," but all sides mounted sophisticated public relations campaigns to plead their case in the newspapers.[1] The presence of a substantial number of journalists at the trial—"a press gallery worthy of a prizefight or a political convention"—was essential to that maneuvering.[2]

For the press, the event promised to sell newspapers and magazines and, in the case of one publisher, to entice listeners to its new radio station. For Science Service, it represented an opportunity to demonstrate the importance of science journalism to mainstream news coverage while also increasing sales. By early July, Davis had lined up orders for special reports from more than a dozen news organizations. The potential of several hundred dollars in additional weekly revenue made it worth sending not one but two staff members to wire stories directly from Dayton.[3]

Watson Davis and Frank Thone had planned to stay at the Hotel Aqua, the unofficial headquarters for visiting journalists. Instead, George Rappleyea arranged for them to be billeted about a mile outside of town in what came to be called Defense Mansion, a grand but deteriorating Victorian house owned by the mining company, which Rappleyea quickly fixed up to accommodate the defense team and its witnesses. That circumstance gave Davis and Thone unusual access to Clarence Darrow and the other attorneys, to the theologians and scientists assisting the defense, and to the celebrities who came to visit. It also completed their transformation from neutral chroniclers to participant-observers.

TELEGRAMS FROM A TRAIN

When Davis returned from California on July 6, he wired Rappleyea for the "names of the scientists to be called on behalf of defense." At 10:45 P.M., Rappleyea replied: "EVERYTHING COMING OFF AS PLANED [*sic*] IF POSSIBLE HELP ME TO SECURE MORE SCIENTISTS TO

TESTIFY TIME IS SHORT QUICK ACTION NECESSARY WIRE ME A LIST THEIR EXPENSES WILL BE PAID WHEN ARE YOU AND SLOSSUM [*sic*] COMING."[4]

Rappleyea then listed eleven possible witnesses (including Henry Fairfield Osborn at the American Museum of Natural History, Columbia University physicist Michael I. Pupin, and Kansas biology professor William M. Goldsmith) with no assurance that any of these men had agreed. Some, in fact, had already told reporters they would *not* travel to Dayton. The scientific side of the defense appeared to be in disarray.

The telegram prompted a flurry of activity in the Science Service offices. Davis and Thone scribbled names of scientists who might be likely witnesses and the journalists quickly sought suggestions from Science Service trustees and from senior men at the National Academy of Sciences.[5] One person gave the names "Pearl, Duggan, Lily, E. B. Wilson, Morgan."[6] Someone else suggested scientists such as W. B. Scott at Princeton University and R. S. Bassler at the Smithsonian Institution. The records show they considered a wide range of possibilities—professors at Harvard, Yale, and Princeton as well as at smaller Midwestern campuses, biologists teaching in Tennessee and Alabama, a number of state geologists, and experts in embryology, physiology, bacteriology, taxonomy, genetics, astronomy, anthropology, geology, and medicine. The rest of the Science Service staff was enlisted to compile information about academic degrees, affiliations, expertise, and where each person could be reached.

Unfortunately, it was already summer.

All around the country, students had graduated, professors had finished teaching, and everyone seemed to be heading off for research trips, expeditions, excavations, or vacations. Some were on their way to Europe; others were ensconced in cottages on remote lakes or trekking in the mountains. It was not easy to locate and to persuade competent and appropriate scientists to change their plans and journey to the South in the middle of July—even for science's current cause célèbre. Nevertheless, the journalists tried.

On July 7, Davis wired Rappleyea a list of sixteen possible experts, with addresses and brief descriptions.[7] Suggestions at that point included Henry Fox (who had lost his job at Mercer University for teaching evolution), botanist John M. Coulter (a "moderator of Presbyterian church and opponent of Bryan"), Charles A. Shull (who had helped to defeat similar legislation in Kentucky), Shailer Mathews ("liberal religious thinker and

Baptist"), Charles H. Judd (psychologist), R. T. Chamberlin ("dean of American geologists [and] originator modern theory earth formation"), and Winterton C. Curtis (a Missouri professor "interested in the humanistic side of science"). Multiple night telegrams were sent from the Science Service offices to nine men, including astronomers Henry Norris Russell at Princeton and Harlow Shapley at Harvard, Yale University paleontologist Richard S. Lull, and science educator Benjamin C. Gruenberg. The text read: "DISTINGUISHED COLLEAGUES OF YOURS HAVE SUGGESTED YOU MIGHT BE WILLING TO COME TO TESTIFY FOR DEFENSE AT DAYTON TENNESSEE NEXT WEEK IN THE CASE OF STATE OF TENNESSEE VERSUS PROFESSOR SCOPES. WE OF THE DEFENSE WOULD BE DELIGHTED TO ADD YOUR AUTHORITY TO OUR POSITION." The telegrams were signed "Clarence Darrow."[8] The responses to these and subsequent wires over the next week attest to the difficulty. R. T. Chamberlin had just returned from the hospital after an operation for appendicitis (but said he would be "ready in a few days" and was willing to testify), Anton J. Carlson was on vacation in Rapid City, V. C. Vaughn was already booked on a lecture tour, Raymond Pearl had "other engagements," astronomers Harlow Shapley and S. A. Mitchell were each in Europe until the end of the summer, and C. C. Little at Woods Hole had "work underway here that cannot be left."[9]

Davis and Thone arranged to travel on the same train with members of the defense team coming from New York. They boarded at Washington's Union Station on the evening of July 8 and "immediately went into a huddle" with attorneys Arthur Garfield Hays and Dudley Field Malone "to plan the scientific phases of the defense" and to select and invite potential witnesses.[10] The journalists brought lists of experts in geology, zoology, anthropology, and other related fields, all screened for availability and willingness.[11] Many invitations were wired en route. Right before boarding, Davis had handed a multiple night letter to Western Union's manager in Washington with a note instructing:

> I am to have a conference with Mr. Malone on board a train en route to Dayton, Tenn., tonight and at that time we shall decide whether the attached night letters are to be sent. I shall send you from the train en route some time after 11 o'clock a wire either canceling or releasing the attached night letters. In the event that they are sent they are to be charged to the account of Science Service.[12]

The journalists continued to assist in acquiring witnesses even once they were in Dayton.[13] Western Union forms were filled out in Davis's handwriting to scientists who had responded positively to the first inquiry ("Thanks for your offer. Will wire you later. Dudley Field Malone") and charged to Cumberland Coal & Iron Company. It had been "quite appropriate for Science Service to have its writers in the press corps covering the event" but, as Davis admitted thirty-five years later, they willingly extended their activities beyond the normal limits: "We went further. We offered to aid in the defense of John T. Scopes. The legal talent marshaled by the American Civil Liberties Union welcomed this aid."[14]

THE SCIENTISTS' SUMMER CAMP

One of the Dayton planners' early concerns was whether there would be sufficient housing for the anticipated crowds of visitors. With the help of the town's Progressive Club, most journalists had been placed at the Hotel Aqua. Arrangements were made for other celebrities, such as Clarence Darrow and his wife, Ruby, to stay in private homes.

Bryan demanded the most elaborate accommodations. Local druggist F. R. Rogers and his wife lived in a large house on Market Street near the courthouse. Rogers told Bryan: "There isn't anything that could please me more than to have you as my guest at home while you are here . . . while not a 'swell' home but a Christian one. I would be glad to place two large rooms at your disposal, and with conveniences."[15] The Rogers family eventually gave up four rooms to the Bryan entourage and, at Bryan's request, arranged for a special cook and maid.[16]

As available space in Dayton dwindled, Rappleyea decided to fix up a derelict mansion on the edge of town for the use of the defense team and its expert witnesses.[17] The Dayton Coal and Iron Company, Ltd., had originally constructed the eighteen-room house in 1884 for the company's local superintendent and visiting stockholders. In recent years, Dayton's teenagers had delighted in holding spooky Halloween parties in the empty rooms, setting candles on the twelve marble mantelpieces and dancing to music played on a portable phonograph.[18]

The imposing but faded yellow building stood on a small rise about a mile out of town, so it offered more privacy for defense preparation. The resourceful Rapp took charge. He ordered mattresses and linens, stripped off peeling paper, painted the walls, brought in minimal furniture, laid down some rush mats, attempted to repair the decrepit water system, and

FIGURE 4.1.
Ova Corvin Rappleyea on the steps of Defense Mansion, Dayton, Tennessee, July 1925. Ova ("Precious") Corvin (1898–1980), a native of Dayton, had married George Washington Rappleyea in 1919. They moved to Dayton sometime in 1922. Precious is shown standing on the steps of Defense Mansion, a dilapidated Victorian house owned by the coal and iron company, which George spruced up to accommodate the defense team and the scientific witnesses. Photograph by Watson Davis. Courtesy of Smithsonian Institution Archives.

arranged to have meals cooked and served on site. Shaded by "majestic trees," the Victorian-style house had "comparative coolness and moderate seclusion" but, despite Rappleyea's ingenuity, only minimal creature comforts.[19] Mencken described it as "an ancient and empty house outside the town limits, . . . crudely furnished with iron cots, spittoons, playing cards and other camp equipment of scientists."[20] On the night before the trial opened, there was no electricity and the plumbing was inadequate: "One and all had been obliged to retire by the soft but inadequate light of candles, and had been awakened by the friendly tap-tap of wood-peckers to a choir of song birds and waterless faucets."[21] Minister Charles Francis Potter, another of the temporary residents, referred to life in the picturesque setting as an adventure dominated by burst pipes and boiled water and interspersed with good conversation and practical jokes.[22]

Both Darrow and Malone camped at Defense Mansion until their wives arrived, and the house served as the main arena for defense strategizing. Davis and Thone lodged there from July 10 to July 22, later reimbursing the ACLU for their expenses.[23] While they were there, Davis and Thone continually juggled two roles—as journalistic observers writing reports for their wire service clients and as informal assistants to the defense. Their involvement was such that they were even listed by several newspapers and in *Science* among the "scientific men" who "actually came to Dayton."[24]

In addition to allowing them to photograph the scientific witnesses, the accidental proximity enabled Davis and Thone to observe the conduct of the defendant more closely. The two journalists had arrived in Dayton already sympathetic to Scopes's cause. By the time they left, they had become defenders of his future; the role they eventually played in the young man's life extended far beyond the normal relationship between reporter and source.

CELEBRITIES

Defense Mansion also proved to be an excellent location for meeting the celebrities who came to Dayton. Historian Jeffrey Moran has noted that, in addition to highlighting differences between North and South, the trial emphasized social and cultural changes affecting the entire country, such as urbanism, the onset of modernity, and the "independence" of women.[25] Although most accounts of the trial focus on the male participants, the presence of a number of strong and accomplished women undoubtedly added spice to the stew. By far the most well known of these was Dudley

FIGURE 4.2.
Publisher E. Haldeman-Julius standing in front of Defense Mansion, Dayton, Tennessee, July 1925. When Emanuel Julius ("E.") married Kansas heiress Marcet Haldeman, they demonstrated their commitment to equality by hyphenating their last name. As the successful publishers of books on science, evolution, and radical politics, they were understandably interested in the attempt to prosecute John Thomas Scopes. The couple drove from their home in Kansas to observe the trial. E. Haldeman-Julius is shown at the front of Defense Mansion, the headquarters for attorneys and scientists who had come to Dayton to assist Scopes. Photograph by Watson Davis. Courtesy of Smithsonian Institution Archives.

Field Malone's wife, Doris Stevens, a prominent Lucy Stoner and executive in the National Women's Party. Stevens had met Malone in 1921 when he offered to defend her and other women jailed for picketing the White House. The statuesque, attractive, and intelligent suffragist apparently scandalized Daytonians with her bobbed hair and smoking and her insistence upon registering at the Hotel Aqua under her maiden name rather than as "Mrs. Malone."

An even more colorful and modern couple drove from Kansas to observe the proceedings. E. Haldeman-Julius, the radical publisher of the Little Blue Books, "was as ardently aggressive in his freethinking as the Fundamentalists in their faith."[26] Born in Philadelphia to Russian Jewish émigrés, Emanuel Julius had devoted his life to promoting communication of liberal ideas.[27] When he moved to Girard, Kansas, in 1915 to write for the socialist periodical *Appeal to Reason*, Emanuel (who preferred to be called simply "E.") met and married Marcet Haldeman, the feminist daughter of one of the town's wealthiest and most prominent families. To demonstrate their commitment to equality, they hyphenated their last name. Like Scripps and the staff of Science Service, the Haldeman-Juliuses were thoroughly committed to the mission of communicating great ideas to the masses. In 1919 they purchased *Appeal to Reason* and its printing plant and began publishing the Little Blue Books series—small, inexpensive books that were primarily new radical texts or reprints of classical literature.

In 1925, the couple were in their mid-thirties. They had published Clarence Darrow's reflections on the naturalist Henri Fabre (*Insects and Men, Instincts and Reason*, 1920) and many other treatises on evolution and would have been well known by reputation to most inhabitants of the Defense Mansion. They had arrived in town just before 3 A.M. on July 10, the opening day of the trial. Marcet Haldeman-Julius described the surreal scene: "Well-lighted and festively bedecked as it was with many banners, not a soul stirred in the streets; a few hounds in front of the stores lay, heads on paws, tails neatly indrawn, eyes closed . . . Dayton was sound asleep."[28] After breakfast, they went out to the Defense Mansion, where they inadvertently drove over and ruptured one of the temporary water pipes.

FIGURE 4.3.
E. and Marcet Haldeman-Julius with Clarence and Ruby Darrow, Girard, Kansas, 1926. Publishers Emanuel ("E.") and Marcet Haldeman-Julius (the couple on the left) had become good friends during the 1920s with attorney Clarence Darrow and his wife Ruby (the couple on the right). This photograph was taken when the Darrows visited the Haldeman-Julius farm in Girard, Kansas, in 1926. Courtesy of Special Collections, Leonard H. Axe Library, Pittsburg State University.

PUBLIC RELATIONS AND THE PRESS

Such celebrities notwithstanding, the turnout of visitors was disappointing. The advance publicity had certainly promised a party. By late June, news articles were emphasizing the entertainment potential, declaring that the "Man vs. Monkey Tilt Has Aspect of Circus" and predicting "sideshows and curious visitors" at the "evolution arena."[29] But, as Westbrook Pegler observed, the old "showman's axiom" held—"if it's free the people don't want it." Crowds never reached even the low estimates, and concessionaires quickly dropped their prices. On the night before the trial, Pegler wrote, "there was plenty of room on Dayton's sidewalks . . . and the lop-eared mules . . . had no need of traffic cops."[30]

Most of the tourists came, in fact, from nearby Tennessee towns: "They came in wagons fitted with settees and chairs and drawn by big-legged horses and small-legged mules. Some came on foot. All were sober-faced, tight-lipped, expressionless, for they were to witness, it seemed to them, a 'battle for the Lord.'"[31] These visitors were not interested in monkey trinkets. At the close of each court session, the farmers piled back into their various conveyances and drove home to milk cows, feed chickens, and finish the day's honest labor.

Fortunately for Dayton's economy, between 100 and 150 reporters and photographers did come to town, although not all stayed to the trial's end.[32] Both sides did their best to attract positive publicity for their cause while disavowing responsibility for the "clatter which . . . has served to confuse the public mind."[33] Everyone but Scopes seemed caught up in the ballyhoo. His father, Thomas, shared his political opinions freely with all reporters during the trial. The judge and the lawyers posed endlessly for photographs, occasionally delaying the trial to do so. Darrow called the judge "a particularly obliging subject" for reporters, even to the extent of suspending proceedings to allow "the boys" in the press to snap photographs or posing on the bench during recess.[34] Local wags characterized the procedure as "Court opens, prayers said, pictures taken, court adjourns."[35] On the weekends, both Darrow and Bryan extended the debate by issuing dueling statements to the press.

The American Telephone & Telegraph Company laid 10.5 miles of temporary wire to speed the transmission of stories; Western Union's twenty-two on-site telegraphers transmitted more than 1.5 million words throughout the course of the trial; and at least another half million went out over a private wire leased by four of the press associations.[36] *New York Times* correspondents alone telegraphed 120,000 words.[37]

Davis absorbed the trial from one of the reporters' tables, seated next to H. L. Mencken, about 12 feet away from the judge and immediately behind the defense team and Scopes.[38] For the young journalist, the trial was

FIGURE 4.4. John Thomas Scopes and his father, Thomas Scopes, July 1925. Thomas Scopes had emigrated to the United States from England in the late nineteenth century; he worked as a railroad machinist in Illinois and Kentucky when young John was growing up. Thomas came to Dayton during the trial to offer support for his son, and the father did not hesitate to share his wit and (occasionally radical) political opinions with visiting reporters. In his autobiography, John described his father as a major influence on his character and "one of the greatest men I have known." Courtesy of Photographic History Collections, National Museum of American History, Smithsonian Institution.

an important opportunity to engage with some of the 1920s' best-known writers and make connections to newspapers likely to purchase his science stories in the future.

The Sun Papers of Baltimore sent not only Mencken but Henry M. Hyde, Frank R. Kent, and cartoonist Edmund Duffy. The artist, known for his uncompromising caricatures of politicians such as Bryan, found ample source material at Dayton's "Big Show."[39] Some of the brightest young political analysts came—Westbrook Pegler, Rollin Lynde Hartt, William K. Hutchinson of the International News Service, Philip Kinsley of the *Chicago Tribune*, magazine writer W. O. "Bill" McGeehan, essayist and Tennessee native Joseph Wood Krutch, and Russell D. Owen of the *New York Times* (who won a Pulitzer Prize five years later). Richard Beamish of the *Philadelphia Inquirer* was appointed as spokesman for the journalists in interactions with the court. Raymond Clapper, chief of the Washington bureau of United Press News syndicate, wrote colorful accounts of the trial's participants, as did Jack Lait, then on the *New York Daily Mirror* staff. Several reporters, such as Michael Williams, founding editor of the Catholic magazine *The Commonweal*, covered the religious angle. The foreign press was well represented and, of course, there was young Nellie Kenyon, the Tennessee cub reporter who had written the first story about Scopes's arrest.[40] Among the oldest members of the visiting press corps was sixty-one-year-old Arthur Brisbane, who had made his reputation as managing editor of Hearst's *New York Journal*, where he perfected the paper's "yellow journalism" approach. Brisbane unabashedly advocated dual roles for newspapers, which conformed well to this trial's sensationalistic potential. The press had a responsibility to enhance democratic discussion, he believed, but newspapers also should provide their readers with occasional entertaining "vaudeville."[41]

Two years after the trial, Davis had occasion to assess how well the press corps had performed. Had journalists been seduced by Bryan or been too sympathetic to the defense? William E. Ritter had agreed to deliver a speech about science journalism at the 1927 AAAS meeting in Nashville, Tennessee, and he asked Davis for "information . . . on the extent to which the newspapers of the country stood for freedom of teaching" at the time of the Scopes trial.[42] Davis dispatched a staff member to the Library of Congress with a list of newspapers, both large and small circulation, from all parts of the country, and then he summarized for Ritter the results of that analysis.[43]

That report, written while memories of the trial were still fresh, remarked upon the considerable disparity between attitudes of individual journalists and the newspapers' editorial positions, especially when location was taken into account. Most reporters who covered the trial "were temperamentally inclined to the pro-evolution side of the case," Davis explained to Ritter, and were therefore more likely to give "breaks" to the defense "so far as the news allowed." Although "support of the defense side of the case does not necessarily mean a philosophical support of evolution," he noted, "the fact is that as a whole the reporters were on the side of science." The editorial attitudes of papers in large cities such as New York, Cleveland, and Chicago were also "unqualifiedly pro evolution" although not necessarily "philosophically serious." Other papers that claimed to be neutral, such as the *Los Angeles Times* and *Toledo News-Bee*, "gave themselves away as anti evolution by praising Bryan or 'the good old religion.'" Southern newspapers, however, "were anti in general, some ardently, others bitterly, and all heaping praise on William Jennings Bryan."[44]

LISTENING IN

Most members of the press corps dealt in print. Two other visitors, however—announcer Quin Ryan and his engineer—symbolized how movie newsreels, radio, and similar technologies were reshaping public attention to national events. The *Chicago Daily Tribune* had persuaded the Rhea County court system to allow the newspaper's radio station WGN to broadcast directly from the courtroom, the first such attempt to capture audio from an ongoing trial.[45] WGN's microphones and announcer were allowed into the courtroom, and in exchange, the station provided amplifiers to allow people outside the courthouse to hear the proceedings.[46] Because of the expense of leasing telephone lines to send the signal back to the Chicago transmitters (estimated to be more than $1,000 per day), WGN eventually decided to broadcast only the trial's more colorful segments, such as the sessions when Darrow or Bryan were likely to speak, and the project was suspended altogether at the end of the second week.[47]

The three microphones, standing "like fence posts" in the courtroom, added to the sensationalism, enhancing the sense of drama and novelty.[48] Although most of the WGN broadcasts simply recorded the lawyers' speeches, Ryan also identified every speaker to his radio audience and, for a half hour before proceedings began, made a special effort to draw listeners' imaginations into the courtroom as he described "every detail

FIGURE 4.5. George Washington Rappleyea, Dayton, Tennessee, July 1925. George Washington Rappleyea (1894–1966) was born in Brooklyn, New York, and attended Brooklyn Polytechnic. Rappleyea had served briefly in the military during World War I and was post commander of the Dayton American Legion in 1925. He was a creative, enterprising, and well-read man who demonstrated considerable expertise in engineering, geology, chemistry, and aeronautics throughout his life. This photograph shows Rappleyea as he escorted visitors on a tour of the remaining mine operations in Dayton during July. Photograph by Frank Thone. Courtesy of Smithsonian Institution Archives.

FIGURE 4.6. Wallace Haggard, Gordon McKenzie, John Randolph Neal, and George Washington Rappleyea, Dayton, Tennessee, July 1925. This photo, published in newspapers under the title "Lounging Around," shows Rappleyea (at right) and Scopes's local attorney John Neal (in dark suit) talking with two members of the prosecution team. Courtesy of Special Collections Library, University of Tennessee, Knoxville.

of the assembling scene, the entrance of each character, the manner, the apparel, the view through the windows, the atmosphere." During the intermissions, he interviewed various visitors and residents, further adding to an aura of entertainment.[49] During the noon recesses, Ryan included "special speeches" by Scopes, Rappleyea, druggist Robinson, several of the attorneys, and even the judge himself.[50]

"OODLES OF LOVE"

Back in Washington, another independent woman, Helen Miles Davis, was struggling to cope with household crises, flat tires, curious kittens, a concrete testing business, and a toddler, but she also found time to field questions from the press. "Your nice wire came this morning just as I was hurrying down stairs to see what the Post had to say about you," she wrote to her husband Watson. "They called up last night for some dope on your career, of which one of the 4 enclosed clippings is the result." Helen also enclosed clippings from other papers, adding her own assessment of which publications were "rather fundamentalist" and which were taking "a decided stand for the evolutionists." Much of the news understandably concerned family:

> I have been spending my time in a variety of ways. Sunday night I changed a tire (rim and all, I'm glad to say), Monday night I went with your Mamma to see Twin Beds at the National (stock). Yesterday I rescued one of Miss Williams' 14 who was suffocating with her head in salmon can. Today I sold 2 TRUNCON Molds, with the prospect of selling 2 more to the same people—Asher Fireproofing Co.—in the near future and a lot more here in the future. . . . Baby [Charlotte Davis] is just fine. I think she will cut her next four teeth soon.

"Don't get typhoid, which, I see has broken out near there," she added, "nor get shot by any wild mountaineers or their moonshine." Like many who were reading the news from Dayton, Helen Davis wished she could be there "to see the excitement for a little while," and she added:

> When are all the people going to start shouting, get the jerks, and all the other symptoms of getting old-fashioned religion?
> I must go get baby some food, she is getting fussy.
> Oodles and oodles of love,
> Helen[51]

CHAPTER 5 RELIGIOUS FEELING

To *Baltimore Sun* writer Frank R. Kent, the "real story" in Dayton was not the celebrity visitors or lawyers but the extent to which religion dominated the lives of residents. "The whole region is saturated in religion," he wrote; religion is a "mode of recreation as well as [a] means of redemption," replacing golf, theater, art, and music as "the fundamental communal factor" of life.[1] To be unaffiliated with a religious group—that is, to refuse to participate in religion—was perceived as tantamount to rejecting the community and its life force. People in the region, Arthur Garfield Hays explained, also took "their particular brand of religion" very seriously: "To the people of Tennessee, salvation is important; they fear the teaching of any doctrine that will destroy faith. It doesn't occur to them that there is something wrong with a faith which is endangered by knowledge, and that fear and threat, fable and irrationality will, among intelligent people, upset faith more surely than intellectual freedom."[2] One episode in particular demonstrates the intensity of these attitudes and their implications for social order.

The Scopes trial had attracted a number of evangelists hoping to preach against evolution, but several well-known liberal ministers had also come to Dayton to support the defense. One of the more brazen of these, Charles Francis Potter, was invited by a local Methodist minister to speak to his congregation. When the sermon's topic was publicized as "Evolution," some church members threatened to lock the sanctuary door, and the young local minister felt compelled to resign. Even though the threat from a single sermon may have been slight, the fear and passion that summer ran deep.

MINISTERS ON THE PROWL

With characteristic nondenominational sarcasm, H. L. Mencken had observed during the trial that, in addition to Bryan's sermonizing at the courthouse, "two Unitarian ministers are prowling around the town looking for a chance to discharge their 'hellish heresies.'"[3] The men he described probably delighted in such characterization. To them, speaking truth to fundamentalists' power represented not heresy but social conviction. They and their wives had traveled to Dayton to observe the trial and offer assistance to the defense, and they, too, were invited to bunk at the rehabilitated mansion on a hill.

Charles Francis Potter had begun his religious life as a fundamentalist Baptist but then switched denominations. In 1919 he had become the minister at the West Side Unitarian Church in New York City and had recently attracted nationwide attention by engaging in lively radio debates about evolution with the Baptist preacher John Roach Straton.

Potter had subsidized his Tennessee trip by agreeing to write articles for several magazines. At age forty, he was at another career crossroads. He was resigning from the New York church and about to embark on an active period of writing and lecturing while serving as executive secretary of Antioch College. A few years later, Potter returned to active ministry as the head of the Universalist Church of the Divine Paternity in New York City and then, in 1929, tiring altogether of the constraints of orthodox religion, founded the First Humanist Society of New York. "I personally ran the whole gamut of change in religion," he later explained in his autobiography, "from extreme orthodoxy to radical liberalism—from fundamentalist Baptist via Modernist Baptist and Unitarian Universalist to Humanist."[4] He and his wife, Clara Cook Potter, had traveled to Dayton on the same train as the defense team and had been met by Rappleyea, who invited them to stay in Defense Mansion.

The other middle-aged theologian, L. M. [Leon Milton] Birkhead, was one of the first major American humanists and, along with Rappleyea, later served on the advisory board of Potter's Humanist Society. Birkhead also had experienced several dynamic religious adjustments. He began his ministerial career as a Methodist, but by 1915, disillusioned with more conservative approaches to religion and driven by social concerns and involvement with the labor movement, he had become a Unitarian, moving to the First Unitarian Church in Wichita, Kansas.[5] In 1925, at age forty, he was minister of All Souls Unitarian Church in Kansas City, Missouri, and had recently attracted press attention by debating the Flying Fundamentalists, a group of antievolution ministers who were traveling around the United States.[6] Birkhead and his wife, Agnes, and their young son drove to Dayton to offer assistance and support to Scopes. While staying in Defense Mansion, Agnes served as a volunteer stenographer and typist for the defense team.

Marcet Haldeman-Julius said that she and her husband first met Birkhead "under one of the great trees on the lawn of the old, and for many years, disused, Mansion House in which the defense lawyers of the Scopes case were domiciled."[7] It was the beginning of a long friendship between the two couples. Birkhead, she wrote, "has a complete absence of pose," and "his utter honesty is not to be questioned, nor his courage," for "he is a naturally frank, candid type."

> Tall, well-built, blue-eyed, fair-haired, open and sunny-faced, he has an engaging, simple, direct manner. An extravert [*sic*], he revels in people. A born mixer, he can see all sides of a question.[8]

Haldeman-Julius concluded that "both the fact of his coming and the manner of it were indicative of the man":

> For he is interested in everything vital and the three Birkheads are the best of chums. Indeed, this minister owes no small part of his rise and freedom from imaginary shackles to the happy fact that his wife, as courageous and debunked as he, has been able to keep joyful step with him in his progress.[9]

A MINISTER'S CONSCIENCE

The first weekend after the trial began, tempers started to rise in Dayton. Emotions spilled out around the edges of politeness and into the humid air.

At the time, Dayton had five Protestant churches, including two Methodist denominations, one of which was slightly (but only slightly) more conservative than the other. John Scopes was a member of the Sampson Bible Class at the more conservative congregation, Dayton Methodist Episcopal Church (South), taught by Judge Ben McKenzie, one of the prosecution lawyers.[10]

The pastor for the other Methodist church, Dayton Methodist Episcopal Church (North), was thirty-five-year-old Howard Gale Byrd. Byrd had been in Dayton for three years, essentially working two ministerial jobs. He was the head of the successful downtown congregation with about a hundred members and of another slightly more conservative one, Five Points Methodist Episcopal Church, about 5 miles away.

The soft-spoken, modest Byrd was generally regarded as one of the best preachers in town and was widely admired for his conscientiousness

FIGURE 5.1.
Howard Gale Byrd, on grounds of Defense Mansion, Dayton, Tennessee, July 1925. Rev. Howard Gale Byrd (1890–1973) had lived in Dayton since 1922. He was pastor of a successful downtown congregation, Dayton Methodist Episcopal Church (North), and another slightly more conservative one, Five Points Methodist Episcopal Church, on the outskirts of town. Byrd resigned from the downtown church during the Scopes trial when members of the congregation objected to a sermon by a visiting minister who had proposed to preach about evolution. Photograph by Watson Davis. Courtesy of Smithsonian Institution Archives.

and hard work. A graduate of the University of Chattanooga and Lincoln Memorial University, the resourceful minister had used his training in carpentry and electrical engineering to work his way through college. In Dayton, he applied the same skills to fix up two run-down sanctuaries. With the help of his parishioners, he rebuilt, repainted, and repaired the Dayton building and then built a parsonage next door for his wife and three children. Byrd was well liked and respected. At the request of Judge Raulston, he had given the opening prayer at the first Scopes court hearing on May 25. He seemed to have been settled comfortably in Dayton—until the summer of 1925.

Byrd's troubles began with Rappleyea. In late June, Rappleyea had been forced to resign as Sunday School superintendent of yet another Methodist church in the area, a post he had held for three years. Although the New Yorker had been liked by most members, a few had objected to his public expressions of support for evolution, especially when those views had been incorporated into a "modern interpretation of the Bible."[11] Rappleyea then joined Byrd's church in town and claimed to the Associated Press on July 1 that Byrd had been the "inspiration" for the Scopes case.[12] When asked to comment on that statement, Byrd disclaimed credit for instigating the trial, but Rappleyea continued to insist that the pastor's "liberal views" had been influential. Byrd (described as "stepping down from a scaffold raised beside the church at Soddy, which he was helping to repaint") did not help his public image among religious conservatives when he told reporters somewhat ingenuously that he was "neither a fundamentalist nor a modernist" but did believe in "progressivism" and "improvement" of life. "True science" and the Bible are not in conflict, he explained to an Associated Press reporter:

> Science deals only with those things that can be analyzed by the five physical senses. . . . Religion is based purely on faith. We should not be concerned so much with whence we came and whither we go, but with the fact that we are here and are here for a purpose. We should leave the world a better place for our having lived in it.[13]

At the end of the trial's first week, plans had been made for William Jennings Bryan to preach at Dayton's Methodist Episcopal Church (South), so Rappleyea suggested to Byrd that Potter be asked to speak to the other Methodist congregation on the grounds that "it seemed very fair to him that a Modernist minister should speak at the same time."[14] Byrd invited

Potter to preach on Sunday night. Then, Rappleyea and Potter raised the stakes. They placed flyers around town announcing that the sermon's title would be "Evolution."

Many of Byrd's parishioners, incensed at the proposed appearance of a "high-faluting preacher from New York," revolted. One group visited the parsonage, and Byrd met with a larger group in the church. On Saturday, July 12, Byrd called off all three of his Sunday services in Dayton and announced to journalists that "I have quit. I have not resigned; I have quit."[15] It was, he believed, a matter of honor because "I invited a brother clergyman to preach in my pulpit tomorrow. I will not withdraw the invitation."[16] He then went to preach to his rural congregation.

Rappleyea and Potter were not finished. Potter informed some journalists that parishioners had "threatened to destroy the church" if the Unitarian spoke: "I do not know whether they meant to tear down the bricks or to break away from the organization."[17] It was, of course, far more likely the latter, but the word "destroy" matched Potter's agenda. He handed out copies of the sermon he had planned to deliver ("The time has come for us to dare to believe that God is big enough to speak to the heart and conscience of every individual through the natural rather than the supernatural.").[18]

The weekend drama featured one more act. Potter was determined to deliver his sermon. So Rappleyea "wangled a reluctant permission from some official for [Potter] to preach on the courthouse lawn that evening."[19] As Potter recalled in his autobiography:

> Platform and chairs were already in place there for an afternoon address by Bryan, so all we had to do was advertise, which was difficult on such short notice. I lettered a sign announcing the meeting and got permission from Mr. Robinson to paste it on the window of his drugstore. In a half hour I went back and found it torn down. So I made another, and we stood near by for a while, meanwhile spreading the news of its having been removed. The news of unfairness spread faster with the news of the meeting than the latter would have circulated without it. . . . That evening I had the majority of the two hundred reporters in my audience.[20]

Potter's talk "was devoted to an exposition of the beliefs of the 'liberals and progressives' in religion."[21]

FIGURE 5.2.
Group on steps of Dayton Methodist Episcopal Church (North), Dayton, Tennessee, July 1925. At far left, holding a coat, is Unitarian minister L. M. Birkhead, who had traveled from Kansas City to observe the Scopes trial. Standing next to him on the first step is Rev. Howard Gale Byrd, pastor of the Dayton church. On the upper steps are (left to right) George Washington Rappleyea, an unidentified man holding one of Byrd's children, and Science Service biology editor Frank Thone. At far right is Charles Francis Potter of New York City, another outspoken liberal minister, holding Byrd's daughter. Photograph by Watson Davis. Courtesy of Smithsonian Institution Archives.

FIGURE 5.3. George Washington Rappleyea, Howard Gale Byrd, and Charles Francis Potter, Dayton, Tennessee, July 1925. The three men are standing at the front of Byrd's church, Dayton Methodist Episcopal Church (North). Rappleyea (left), who had only recently joined Byrd's congregation, had persuaded the minister to invite Potter to give a sermon. Rappleyea argued that because William Jennings Bryan was scheduled to preach that same Sunday at Dayton Methodist Episcopal Church (South), it would be only "fair" if a liberal minister such as Potter spoke at Byrd's church. Photograph by Watson Davis. Courtesy of Smithsonian Institution Archives.

FIGURE 5.4.
Howard Gale Byrd and Charles Francis Potter, with Byrd's children, Dayton, Tennessee, July 1925. Howard Gale Byrd (left) had just resigned as minister of Dayton Methodist Episcopal Church (North). Byrd and visiting minister Charles Francis Potter are standing in front of Byrd's parsonage. Photograph by Watson Davis. Courtesy of Smithsonian Institution Archives.

FIGURE 5.5. Temporary privies built behind Rhea County Courthouse, Dayton, Tennessee, July 1925. The Dayton townspeople worked diligently to accommodate the anticipated crowds of visitors for the trial of John Thomas Scopes, including providing temporary outhouses near the courthouse. This structure, designated "For Ladies," had been draped with a "Read Your Bible" sign similar to the one hung on the courthouse itself. Frank Thone and Watson Davis took several photographs of this coincidence of mortal and spiritual needs. When Thone returned home to Washington, he had copies of the photograph printed as a postcard and mailed them to friends around the country. Photograph by Frank Thone. Courtesy of Smithsonian Institution Archives.

SPELLING IT OUT

Byrd appears to have kept his sense of humor throughout the episode, even if he was out of work. The day after his resignation, he emphasized to the *Chicago Tribune* that "My name is spelled with a 'y.' . . . Evolution has carried me beyond the feathered species."[22] Fortunately, Davis preserved images of the actors in this revealing sideshow. In one photograph (Figure 5.2), arranged on the steps of Byrd's church stand the principal players: the ever-present Rappleyea, the journalist Frank Thone, the two visiting ministers bent on a little rabble-rousing, and the gentle southern preacher, doe-eyed but resolute in the face of the ideological passions swirling around him.

CHAPTER 6 THE SCIENTISTS COME TO TOWN

During the trial's second week, scientists descended on Dayton. The men who had agreed to serve as expert witnesses anticipated a cerebral battle. They had expected to engage in energetic but dignified courtroom debates in which reason and enlightenment would prevail and would "reveal to an unconverted jury, judicial and public, the evidences for evolution."[1] Instead, the visiting scientists confronted stalemate, stalled action, and judicial obstructionism. In the end, only one scientist made it to the stand.

While everyone waited, Watson Davis and Frank Thone took advantage of the circumstances to get to know the scientific witnesses lodged at Defense Mansion (many of whom later became sources or subjects of Science Service news stories) and to photograph them. Thone's portraits in particular captured well the scientists' mood of confident defiance. These were not the usual formal portraits of the day, such as those illustrating scientific textbooks or magazine articles. The scientists in Dayton stood with sleeves rolled up and arms crossed, eager for a fight.

Detached from the symbols of their trade, without the standard trappings of microscopes, beakers, test tubes, or flasks, and shed of the white laboratory coats draped so ubiquitously onto popular culture's scientists, these accomplished researchers were transformed into human beings. Posed outdoors against sylvan backgrounds of dappled shade, they might have been Scopes's favorite uncles, cousins, or friends, lingering after a Sunday dinner. Instead, they were world-renowned experts gathered to defend the teacher's freedom to discuss evolution.

RELUCTANT WITNESSES

Assembling the group had been far from easy. Within a few days of the trial's opening, the defense had still not secured all its expert witnesses. The correspondence and multiple telegrams generated by Davis, Thone, and Rappleyea between July 8 and 15 attest to the difficulty of arranging the participation of suitable witnesses when so many scientists were abroad, conducting field research, or sequestered somewhere on vacation.[2] The obstacles, however, did not stem only from scheduling or family obligations. Many scientists—including some of the most well known names in the United States—apparently found the trial's potential sensationalism intimidating. As Philip Kinsley reported that week: "The defense is having trouble in getting prominent men to come . . . to testify on the side of evolution, even theistic evolution. These gentle souls from the cloistered schools and churches fear to face what they imagine is a fierce rabble down here, fear the position which Bryan will force them into if he can do so."[3] Attacking Bryan in print was one thing. Facing him in the public setting of a Tennessee courtroom smacked of spectacle and undesirable publicity.

Nevertheless, the importance of the cause, the reputation of Clarence Darrow, and the perseverance of the defense team and its supporters prevailed. During the week of July 13, an impressive group of men straggled into Dayton, many of them with strong ties (either as graduates or faculty) to the University of Chicago, Thone's alma mater.

THE SCIENTIFIC WITNESSES

Sometime that week, Davis wired to the Science Service offices a list of thirteen "scientific witnesses to be called" as experts, to be published in the *Daily News Bulletin* mailed to their clients.[4] Eventually, only one of those men was allowed to testify in open court.

On Wednesday, July 15, court opened, as usual, with a prayer. That day, the petitioner (under a rotation among denominations that had been arranged after considerable negotiation) was Charles Francis Potter, who asked the Lord to lift up the hearts of all in the courtroom so that "we may seek Thy truth"—a goal Potter personally defined as belief in evolution.[5] The rest of the morning was consumed with arguments over a defense motion to quash the indictment and then with Judge Raulston's lengthy opinion ruling in favor of the state. That afternoon, county school superintendent Walter White and school board president F. E. Robinson testified, as did fourteen-year-old Howard Morgan, a student from Scopes's class. Around 4 P.M., the prosecution rested, and Darrow plunged forward with the case for the defense.[6]

The first scientific witness, fifty-seven-year-old invertebrate zoologist Maynard Mayo Metcalf, had arrived in Dayton the previous day. He took the stand and responded to Darrow's questioning about evolution for less than an hour. Metcalf was a good choice for opening witness; calm and precise, he had strong credentials in both science and religion. A native of Ohio, Metcalf had taken graduate degrees from both Johns Hopkins University and Oberlin College and had been teaching for several years in the Oberlin zoology department. In 1925, he was serving as chairman of the National Research Council's Division of Biology and Agriculture and was about to become a professor of zoology at Hopkins. An active Congregationalist, he frequently taught Bible classes at his church.

Thone and Davis reported that Metcalf's explanations were so clear that they "caused the courtroom audience to lean forward." Even William Jennings Bryan seemed to attend to the scientist's remarks. Metcalf carefully and conscientiously distinguished "between the facts of evolution and the numerous theories of how evolution came about," but in closing, he emphasized that "no normal man can hold any doubts of the facts of evolution."[7]

Such precision and caution were of little avail. On Thursday, the prosecutors began a series of maneuvers to block the testimony of the remaining defense experts, culminating on Friday, July 17, with the judge's ruling to exclude any more testimony in open court, although he did allow formal written statements to be submitted so they could form part of the official record for the expected appeal.[8]

One of the youngest of the scientific experts who came to Dayton (and one of those permitted to submit an affidavit) was thirty-seven-year-old Kirtley F. Mather. He had just been named chairman of the geology department at Harvard University. Like Metcalf, he also possessed a useful combination of scientific credentials, expertise, and religious affiliation. After graduating from a Baptist institution, Denison College, Mather had returned to his hometown to earn a Ph.D. at the University of Chicago. He happened to be familiar with the geology of Tennessee because he had conducted research there; in addition to an impressive scientific record, he was active in the Baptist church.[9]

Mather had been tracked down in Whycocomagh, Nova Scotia, but the distance apparently did not deter him. On July 8, he wired back asking when the trial date was. On July 14, he wired his travel plans from Nova Scotia, saying that he planned to go first to Massachusetts and then to travel to New York City on Thursday morning, July 16. His train south would leave Pennsylvania Station Thursday evening, with scheduled arrival in Chattanooga on Friday night at 11:30 P.M.

The University of Chicago was well represented at Dayton, with not

FIGURE 6.1.
Maynard M. Metcalf, ca. 1925. Zoologist Maynard Mayo Metcalf (1868–1940) was the only scientist allowed to testify on the stand at the Scopes trial. Trained at Oberlin College and Johns Hopkins University, Metcalf had just been appointed research associate and professor of zoology at Johns Hopkins after years of teaching at Oberlin. He was also a devout Christian and active in the Congregationalist church. This photograph was mailed by Science Service to its clients to accompany an article about the scientists expected to testify for the defense. Courtesy of Smithsonian Institution Archives.

only graduates such as Mather and Thone but also three faculty members. The dean of the university's School of Education, Charles Hubbard Judd, arrived in Dayton Tuesday afternoon. The fifty-two-year-old psychologist had been born in British-controlled India and had come to the United States as a child in 1879. Educated at Wesleyan University, University of Leipzig, Yale University, and Miami University, he had become professor of education at Chicago in 1909, the same year he was president of the American Psychological Association.

Zoologist Horatio Hackett Newman was the only one of the expert witnesses who had been born in the South. A native of Seale, Alabama, the forty-nine-year-old had earned a Ph.D. from the University of Chicago in 1905 and had been professor of zoology there since 1917. Like many of the other witnesses, Newman had considerable administrative experience, having only recently stepped down as dean in the College of Sciences. The biographical information accompanying his formal statement emphasized that he was "among the earliest in this country to organize large classes in various universities for the study of evolution and heredity."[10] Years later, he claimed that the evolution controversy had had a positive effect on science because it had "greatly enhanced" popular interest in biology and increased the number of students electing to study genetics.[11]

Newman had traveled to Dayton that week with the third Chicago faculty member, anthropologist Fay-Cooper Cole. Cole had been educated at the University of Southern California, Northwestern University, and Columbia University. For many years he worked as an ethnologist at the Field Museum, conducting three separate expeditions over five and a half years to such places as the Philippines, Borneo, Java, Sumatra, and the Malay Peninsula. In 1925, at age forty-three, he had just been appointed associate professor of anthropology at the University of Chicago.

Like Judd, fifty-year-old agricultural scientist Jacob Goodale Lipman, dean and director of the New Jersey Agricultural Experiment Station at Rutgers University, had been part of the late-nineteenth-century waves of immigration into the United States. Born in Friedrichstadt, Russia, to a prominent Jewish family, Lipman had attended school in New Jersey in the 1890s and earned a Ph.D. at Cornell University in 1903. By 1925, Lipman was a distinguished professor of agriculture with an international reputation in soil science, and yet he had agreed to come to Dayton immediately upon receiving the invitation, asking only to be notified "whether you need me and when."

FIGURE 6.2.
Fay-Cooper Cole, Dayton, Tennessee, July 1925. Anthropologist Fay-Cooper Cole (1881–1961) had just been appointed associate professor at the University of Chicago after years of conducting fieldwork in the Malay Peninsula, Sumatra, Java, Borneo, and the Americas. He was the first anthropologist in the sociology department at Chicago, and when a separate department of anthropology was created, he served as its chairman from 1929 to 1947. Photograph by Frank Thone. Courtesy of Smithsonian Institution Archives.

FIGURE 6.3.
Jacob G. Lipman, Dayton, Tennessee, July 1925. Jacob Goodale Lipman (1874–1939) was professor and dean of agriculture and the director of the New Jersey Agricultural Experiment Station at Rutgers University. An internationally known expert on soil science, Lipman had been born in Friedrichstadt, Russia, and educated in the United States, receiving his Ph.D. from Cornell University in 1903. Although the prosecution prevented Lipman and most of the defense scientists from testifying in open court, their affidavits were read into the record on Monday, July 20, 1925. Photograph by Frank Thone. Courtesy of Smithsonian Institution Archives.

Another of the geology experts, thirty-five-year-old Wilbur A. Nelson, had been state geologist of Tennessee since 1918 and was scheduled to move to Virginia in September to join the faculty of the University of Virginia and to head its geology department. Nelson had been educated at Vanderbilt and Stanford Universities. He was president of the Monteagle Sunday School Assembly (the interdenominational Chautauqua and summer resort founded forty-six years before on the Cumberland Plateau above Dayton), but he embraced the evolution displayed in the rocks he studied and was therefore, Thone remarked, "a lonely official in a fundamentalist state."[12]

Although the expenses of some witnesses were eventually reimbursed by the ACLU, a few paid their own way. The forty-nine-year-old Winterton Conway Curtis, professor of zoology at the University of Missouri, had been born in Maine, earned a Ph.D. from Johns Hopkins University in 1901, immediately accepted a job at Missouri, and had been there ever since. Author of many scientific texts, Curtis had also recently turned his attention to the social aspects of science, in such works as *Science and Human Affairs from the Standpoint of Biology*.[13]

For the scientists, the trip to Dayton had been far from straightforward, either professionally or physically. Their willingness to make the arduous journey gave evidence of their commitment to the cause. Curtis, for example, had asked in his July 6 telegram to be given notice "well in advance" of when he would be needed because the trip from Columbia, Missouri, to Dayton would take from noon on one day until 6 P.M. the next. He signaled that he would be willing to stay for a day before and after his testimony if accommodations could be secured and he even offered to come "at own expense."[14]

Several other scientists traveled to Dayton and stayed at Defense Mansion but were apparently not allowed either to testify or submit affidavits on Scopes's behalf. William A. Kepner, professor of biology at the University of Virginia, had been teaching that summer, but he rearranged his schedule and arrived in Dayton around 6 P.M. on July 16. William Marion Goldsmith, a thirty-seven-year-old professor of biology, had traveled from Southwestern College, a Methodist institution in Winfield, Kansas. Goldsmith's research centered on evolution, heredity, and genetics, and soon after joining the Southwestern faculty in 1920, he had begun to express his strong religious views in print, attempting to reconcile evolutionary

FIGURE 6.4.
Wilbur A. Nelson, Dayton, Tennessee, July 1925. At the time of the Scopes trial, Wilbur A. Nelson (1890–1969) was state geologist of Tennessee but soon thereafter he moved to Virginia to become a professor at the University of Virginia and was appointed that state's official geologist. Nelson had recently served as president of the Monteagle Sunday School Assembly at the Chautauqua summer camp atop the Cumberland Plateau near Dayton. Photograph by Frank Thone. Courtesy of Smithsonian Institution Archives.

FIGURE 6.5. Winterton C. Curtis, Dayton, Tennessee, July 1925. Zoologist Winterton Conway Curtis (1875–1969) received his Ph.D. from Johns Hopkins University in 1901 and immediately joined the faculty of the University of Missouri, where he taught for the rest of his career. In addition to his scientific work, Curtis was interested in the humanistic aspects of science. During the early 1930s, he had an opportunity to become director of Science Service but declined the job. Photograph by Frank Thone. Courtesy of Smithsonian Institution Archives.

FIGURE 6.6.
William A. Kepner, Dayton, Tennessee, July 1925. Kepner was a professor of biology at the University of Virginia. He had been invited by the defense to come to Dayton and had rearranged his teaching schedule to appear, but in the end, even his written testimony as an expert witness was not submitted to the court. Photograph by Frank Thone. Courtesy of Smithsonian Institution Archives.

FIGURE 6.7. Scientists gathered at Defense Mansion, Dayton, Tennessee, July 1925. In the back row, left to right, are Horatio Hackett Newman (University of Chicago), Maynard Mayo Metcalf (Johns Hopkins University), Fay-Cooper Cole (University of Chicago), and Jacob Goodale Lipman (Rutgers University). In the front row, left to right, are Winterton Conway Curtis (University of Missouri), Wilbur A. Nelson (state geologist of Tennessee), and William Marion Goldsmith (Southwestern University, Winfield, Kansas). Photograph by Watson Davis. Courtesy of Smithsonian Institution Archives.

theory with the Bible.[15] In 1924, two of his books—*Evolution and Christianity* and the subsequent *Evolution or Christianity, God or Darwin?*—challenged the arguments of the antievolutionists and brought Goldsmith to national attention, resulting in unsuccessful attempts to censure him during the 1924–1925 academic year.[16]

The defense's religious experts were also not allowed to testify on the stand. Formal affidavits were submitted by two local ministers—Rev. Walter C. Whittaker, the rector of St. John's Episcopal Church in Knoxville, and Dr. Herbert E. Murkett, pastor of First Methodist Church in Chattanooga—and by Rabbi Herman Rosenwasser of San Francisco, who had been advising the defense on the differences in biblical texts. Like many of the experts, the forty-six-year-old Rosenwasser exemplified the depth of life-experience and education brought to Scopes's aid that summer. After emigrating from Hungary in 1893, Rosenwasser had studied at Hebrew Union College, earned a degree in semantics and philosophy at Western Reserve University, and been ordained as a rabbi. An additional group of at least sixteen noted scientists and biblical experts had signaled their willingness to assist the defense. For example, Ellsworth Faris, chairman of the department of sociology and anthropology at the University of Chicago and a native of Salem County, Tennessee, helped to coordinate participation by University of Chicago faculty and had been ready to go to Dayton if necessary.[17] At Darrow's request, Shailer Mathews, dean of the University of Chicago School of Divinity and president of the Federal Council of Churches of Christ in America, sent a formal statement for the court record. Mathews was one of the leading scholarly proponents of "higher criticism"—the use of historical literary tools to interpret the Bible—and in such works as *The Faith of Modernism* and *The Contributions of Science to Religion*, he resoundingly declared that "religion had nothing to fear from advances in science."[18]

DEFENDING LIBERTY

In one of their news dispatches that week, Davis and Thone observed that when the scientific experts arrived, most were still not "convinced of the dangers of fundamentalism" because they perceived the trial as either "an impossible dream" or "a nightmare that will disappear in the morning."[19] By Friday, July 17, after Judge Raulston sided with the prosecution and kept the remaining defense witnesses from the stand, the scientists were all believers.

FIGURE 6.8. Scientists, theologians, journalists, and defense lawyers, sitting on the steps of Defense Mansion, Dayton, Tennessee, July 1925. In back row, left to right, are E. Haldeman-Julius, George Washington Rappleyea, Frank Thone, and Watson Davis. Standing at left is Judge John R. Neal, the Knoxville attorney who first represented John Thomas Scopes. Standing at right is Neal's brother-in-law, W. E. Wheelock, the superintendent of the Southern Pacific Railroad Terminal in Chattanooga. Sitting in the middle row, left to right, are Maynard Mayo Metcalf, Charles Francis Potter, Mr. McCleskey (a court stenographer hired by Darrow), William A. Kepner, and Arthur Garfield Hays. Sitting in the front row, left to right, are Wilbur A. Nelson, Fay-Cooper Cole, Winterton Conway Curtis, Horatio Hackett Newman, and Jacob Goodale Lipman. Courtesy of Smithsonian Institution Archives.

FIGURE 6.9. Defense attorneys, journalists, and expert witnesses assembled for the Scopes trial, Dayton, Tennessee, July 1925. The experts gave informal seminars at Defense Mansion throughout the week, and this photograph shows one such talk. Standing second from left is Frank Thone; then from left to right are William A. Kepner, George Washington Rappleyea, Wilbur A. Nelson, Watson Davis (barely visible), Charles Francis Potter, and an unidentified man who is probably Rabbi Herman Rosenwasser of San Francisco, a biblical expert who assisted the defense. On the other side of the chart, left to right, are Arthur Garfield Hays, John R. Neal, John Thomas Scopes, and Dudley Field Malone. Courtesy of Special Collections Library, University of Tennessee, Knoxville.

From the outset, the prosecution had sought to deny science a hearing, had attempted to exclude scientific evidence and the facts of evolution from being presented to the court. The defense, intent on turning the courtroom into "the world's largest schoolroom," had been equally determined to explain the principles of evolution to the jurors and the public. The judge's ruling signaled that "legalistic intolerance" (which Davis and Thone called "fundamentalism by law, stifling freedom of thought") was possible.[20]

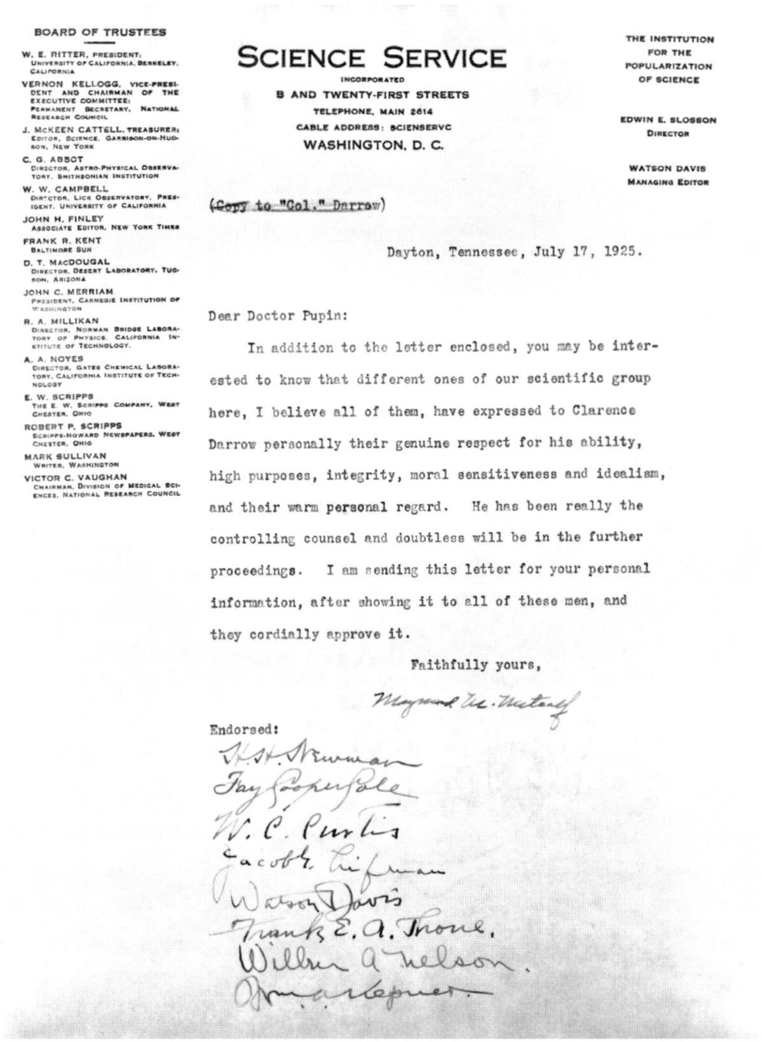

SCIENCE SERVICE

INCORPORATED
B AND TWENTY-FIRST STREETS
TELEPHONE, MAIN 2614
CABLE ADDRESS: SCIENSERVC
WASHINGTON, D. C.

THE INSTITUTION FOR THE POPULARIZATION OF SCIENCE

EDWIN E. SLOSSON
DIRECTOR

WATSON DAVIS
MANAGING EDITOR

~~(Copy to "Col." Darrow)~~

Dayton, Tennessee, July 17, 1925.

Dear Doctor Pupin:

In addition to the letter enclosed, you may be interested to know that different ones of our scientific group here, I believe all of them, have expressed to Clarence Darrow personally their genuine respect for his ability, high purposes, integrity, moral sensitiveness and idealism, and their warm personal regard. He has been really the controlling counsel and doubtless will be in the further proceedings. I am sending this letter for your personal information, after showing it to all of these men, and they cordially approve it.

Faithfully yours,

Maynard M. Metcalf

Endorsed:

H. H. Newman
Fay-Cooper Cole
W. C. Curtis
Jacob G. Lipman
Watson Davis
Frank E. A. Thone.
Wilbur A. Nelson.
Wm. A. Kepner.

FIGURE 6.10. Letter to Michael Pupin from the "scientists at Dayton," July 17, 1925. In signing this letter to physicist Michael I. Pupin (who was president of the American Association for the Advancement of Science) and by using Science Service stationery for the correspondence, Watson Davis and Frank Thone abandoned all pretense of journalistic neutrality and declared themselves part of the "scientific group" at Dayton. Signers (in order) are Maynard M. Metcalf, H. H. Newman, Fay-Cooper Cole, W. C. Curtis, Jacob G. Lipman, Watson Davis, Frank E. A. Thone, Wilbur A. Nelson, and Wm. A. Kepner. Courtesy of Library of Congress.

Many of the scientists at Dayton also apparently changed their opinions of Clarence Darrow, finding him to be "not at all the ogre that the press makes him out to be" and seeing that "underneath his rather forbidding shell there lies a character of real solid worth and even gentleness."[21] "Although Darrow is not a scientist," Thone wrote to one friend, "he surprised all of us by the range and accuracy of his scientific knowledge, and he completely confused Mr. Bryan on both scientific and biblical grounds."[22]

Upon returning to Defense Mansion that Friday night, the scientists and the science journalists composed a testament acknowledging Darrow's contribution to science's fight for freedom. The letter, drafted by Thone and typed on Science Service stationery, was directed to Michael Pupin, president of the American Association for the Advancement of Science, from Metcalf, who explained that he was writing on behalf of the "scientific group" in Dayton in order to praise Darrow's "ability, high purposes, integrity, moral sensitiveness and idealism."[23] The letter was signed by Metcalf, Newman, Cole, Curtis, Lipman, Nelson, Kepner, Davis, and Thone.

After the dust had settled and he had returned home, Thone expressed his regrets that "all the scientists of the country" could not have been in Dayton, "for there was much instruction to be had for all of us concerning the extent and reality of the danger, both to liberty of science and liberty of religion, in that strange revival of Old Testament theocracy which Mr. Darrow was our leader in combating."[24]

Despite the heat and lack of modern conveniences, it must have been a rollicking time at Defense Mansion. "After its long sleep," Marcet Haldeman-Julius wrote, the "roomy old house" had "come to life" and briefly "presented a warm, livable aspect," filled with good conversation, intense debate, and intellectual camaraderie.[25] As Thone wrote: "All day long and far into the night the rumble of scientific discussion and laughter issues forth from Defense Mansion, that pleasant old house on the outskirts of Dayton that has become the headquarters for the defenders of science, religion and freedom."[26]

CHAPTER 7 SUNDAY EXCURSIONS

Even when the cool night breezes eventually brushed their antediluvian quarters, heaven may have seemed far away to the temporary inhabitants of Defense Mansion. Encamped in nearby Morgantown Hollow, however, other visitors to Dayton, followers of the Holiness movement, were certain that they were drawing closer to heaven on earth.

The popular press in the 1920s had labeled the more demonstrative Pentecostal sects as "holy rollers" because worshippers would shout, roll on the ground, and sometimes speak in tongues. Throughout the Scopes trial, one such group held nightly camp meetings near Dayton. Their exotic practices drew caustic comments from H. L. Mencken and their fire-lit worship services afforded easy entertainment to many of the other reporters and visitors.

On Sunday afternoon, July 19, the same Holiness group held an open-air baptism service at a nearby rocky creek. Standing on the opposite bank, Watson Davis took a sequence of extraordinary photographs of the minister and his flock. The complex, evocative images recorded a scene that was common throughout the rural South yet rarely documented by contemporary photographers.

In Dayton that summer, the insider and outsider roles twisted and turned like leaves in the wind. At the baptism, sophisticated easterners such as Davis became observers with their noses against a cultural window, disadvantaged by their inadequate understanding of the sect and its customs. The Holiness worshippers, although disdained by members of the town's churches, stood privileged and serene within the center of their faith, unconcerned about the world's opinions, dismissive of modernity and the evolutionists' evidence, and joyful in their search for personal salvation.

IN SEARCH OF HOLINESS

The East Tennessee sect led by Preacher Joe Leffew was part of a substantial religious movement related to but separate from William Jennings

Bryan's mainstream Christian supporters.[1] During the first decade of the twentieth century, thousands of American Protestants had broken away from traditional denominations to form their own churches or independent congregations as part of what is called the Holiness movement. These evangelical groups, some in independent sects led by ministers such as Leffew, others in loosely affiliated urban and rural churches numbering in the tens of thousands, had grown substantially by 1925. All traced their roots to John Wesley and the notion of "entire sanctification." They cultivated religious experiences that were highly individualistic and highly subjective, based on common themes of personal salvation, Holy Ghost baptism, divine healing, and belief in an imminent millennium.[2] Frank Thone described the Dayton group's theology as "Early Wesleyan," their hymn-singing as "Early Reformation," and their phraseology as "Elizabethan," but he added that "their most characteristic expressions" were their "rigid ecstasies, . . . spasmodic gestures and dance steps, their speaking with strange tongues," which he likened to the "revels of Dionysus."[3]

Such worship practices had prompted many observers (and especially the press) to apply the name "holy rollers" somewhat indiscriminately (and derisively) to all Holiness groups because worshippers were encouraged to shriek, shout, sing, speak in tongues (glossolalia), and roll on the ground to express their spirituality. Outsiders who watched such groups during the 1920s "sometimes left bemused, sometimes incredulous, but rarely indifferent," Grant Wacker has noted in his history of American Pentecostals, and the news reporters tended to fasten on the "turbid emotionalism," "apparent pandemonium," and the "chaotic and deafening" worship. And yet, what appeared to be chaos was actually, Wacker pointed out, a finely tuned oscillation between "antistructural and structural impulses," that is, it was "planned spontaneity" designed to enhance spiritual experience.[4]

The presence of the Leffew group in Dayton during the time of the trial was also not accidental. Whatever the variations among the congregations, the Holiness movement in general was uniformly opposed to Darwinism. Thone mentioned that a member of the congregation told him they had "a general headquarters" in nearby North Carolina, opposed "such modern heresies" as evolution and even Copernican astronomy, and insisted that the earth was flat.[5] Some of these groups accepted more science than others, Ronald Numbers has pointed out, but all drew the line at human evolution: "Evolution and Holy Ghost religion simply did not mix."[6]

Holy Ghost worship also followed divine rather than mortal rules. These were not Sunday-morning Christians. They worshipped any time, any place. Small, poor congregations such as Leffew's often did not have their own church building. During summertime camp meetings, the natural world became their sanctuary. Frank Thone described ceremonies "performed in a starlit woodland cathedral" only a few miles from the courthouse.[7] "A group of them camped outside Dayton during the trial and many of us were wont to observe their weird ceremonies . . . held at night under huge trees from which swing oil flares," recalled Arthur Garfield Hays.[8] "The Holy Rollers of Dayton have no church," Allene M. Sumner wrote in the *Nation*; instead, they use an "arbor church," convening between two great elm trees, the limbs overarching the group, and with whippoorwill and bobcat as background music.[9]

The services blended sound and action. "As they grew excited and shouted and sang and twitched and twirled," Clarence Darrow wrote in his autobiography,

> the people crowded closer around them in curiosity and wonder. Now and then someone would sidle forward from the dark woods and, seeming to be seized with some inspiration, would rush in amidst the other performers and dance and squirm and shout and stutter with such vigor and contortions that the regulars were put to shame for their mild form of worship.[10]

The sensitive descriptions of Thone, Hays, and Darrow contrasted sharply with Mencken's acerbic commentary. The columnist held little respect for any religion. To Mencken, the group offered convenient local color and epitomized the cultural prejudices fueling the trial. Unforgiving portraits of the "howl of holiness" peppered many of his dispatches.[11] The correspondent for a British publication, the *Spectator*, wrote similarly unflattering and sarcastic descriptions of the Holiness group, calling them people who "sit about under trees and howl" and engaged in "no better than Voodoo worship."[12]

Mencken's colleague on the *Baltimore Sun*, Frank R. Kent, offered more temperate observations. He watched a hundred members of the sect for three hours as they "prayed and sang violently until many of them were lying on the ground in a horrible state of physical exhaustion and mental collapse" and he then compared their religious convictions to those of a well-dressed crowd in Dayton that had been meeting that same night. Kent argued that the less sophisticated rural participants were no less sincere in their faith than the sanctimonious Daytonians who debated "various points of Biblical interpretation until long after midnight."[13]

To Thone, the "worshippers in a pagan setting" represented a powerful reaction to social and cultural change. Religious groups and isolated communities such as these were "beginning to feel the pressure of exterminating forces penetrating into their remote shelters." After watching the group one night, he drew on his botany background for a striking analogy. They were, he wrote, much like the Torrey pines in California—surviving because they are "sheltered by their isolation" and representing "both the last stand and hangover of a past age."[14] Botanists call such survivors "relict endemics," he explained, and the Holiness groups were likewise being pushed out by science and modernism yet continuing to resist the "invasion of their endemism by a strange idea."

> There are sometimes endemics within endemics. When the Torrey pine was a dominant tree over thousands of miles of Southwestern hills the California sycamore flourished lordly in the valleys. It is in the valleys still and extends over a much larger domain than does the pine, but in sadly diminishing numbers. It is also a dwindling if not a vanishing species.
>
> There are hundreds of thousands of people in the United States in whose minds there is an irreconcilable division of which they are not even conscious. They no longer believe the Biblical astronomy made popular by Milton, but they do accept literally the words of the Biblical account of the creation of the earth.[15]

Darrow and Hays also appear to have admired the group even as they marveled at the participants' exotic practices. Darrow told reporters that he found the Holy Rollers to be far more estimable than Bryan because they did not seek to impose their beliefs on others: "They at least don't demand that any one join them. They leave other people alone, and their church is free."[16] Many years later, Hays commented that he had "never observed more sincere and deeply religious people than the Holy Rollers."[17]

ON THE OPPOSITE BANK

As the trial rolled on, the temperature rose in and out of the courtroom. Everyone welcomed the weekend recess. Dayton's hospitality still overflowed, but no one could simply push the weather away with a smile. The hotels and rooming houses were crowded and uncomfortable. By the time that court adjourned on Friday, July 17, many participants were eager to escape elsewhere. Some reporters, including Mencken, headed home, convinced that the trial held no more suspense and little news. Even William Jennings Bryan Jr., who had been assisting his father and the prosecution, returned to his law practice in California.

The defense team had been prevented from putting all but one of its expert witnesses on the stand. Now they planned to spend the weekend preparing the remaining experts' affidavits so that the texts could be entered into the court record. The *New York Times* complained that, as "typewriters clicked merrily in the Mansion," the so-called "'duel to the death' has become a battle of statements."[18] Even Bryan seemed enervated. On occasion, his eyes glittered and he sparred with Darrow or Malone in the courtroom, but the displays of legal fireworks were fleeting. Instead of the much-anticipated oratorical clashes, the proceedings had been predominantly cerebral, intellectual, and tedious to watch. The defense was already preparing to appeal a conviction.

That Sunday, Darrow and Malone went to Chattanooga—Darrow to lecture about Tolstoy to the Young Men's Hebrew Association, and Malone and his wife, Doris Stevens, to visit friends.[19] Arthur Garfield Hays stayed behind to help the witnesses prepare their statements. The defense also planned, with the assistance of Davis, Thone, and their organization, to mimeograph those texts so they could be distributed to the press on Monday for immediate publication and maximum publicity.

In the afternoon, Davis, Thone, Hays, and various other Defense Mansion residents took a break and went to see the open-air baptism held by Leffew's group. To natives of the southern United States, the scene would have seemed familiar. Well into the 1950s, immersions in muddy streams could still be glimpsed every Sunday from nearby highways. To the easterners, however, the setting undoubtedly seemed alien and primitive. As Hays recalled: "The worshipers sat, kneeled or stood at the side of a stony creek bed. Children were throwing pebbles into the water and were happily dancing around. The adults were tense and nervous, keyed up for an emotional spree."[20]

Standing on the opposite bank, Davis took eight photographs in sequence that day, recording the progress of the activities. The Holiness group assembles on rocks near the water and next to a ramshackle cabin. At one point, the participants seem to be discouraging two onlookers (perhaps reporters) from coming too close, and the intruders run away.

FIGURE 7.1. Worshippers assembled for a baptism in a stream near Dayton, Tennessee, July 19, 1925. During the weeks of the Scopes trial, an independent group of Holiness movement followers, who rejected the theory of evolution and other expressions of modernism, camped near Dayton and held services every night. Many of the visiting journalists, scientists, and attorneys had observed the group's nighttime services. On Sunday, July 19, 1925, some of the residents of Defense Mansion, including Arthur Garfield Hays, Frank Thone, and Watson Davis, watched an outdoor baptism service being held in a nearby creek. Davis took eight photographs of the events. This image shows the group assembling on the bank and appearing to chase away two interlopers. Photograph by Watson Davis. Courtesy of Smithsonian Institution Archives.

FIGURE 7.2. Worshippers assembled for a baptism in a stream near Dayton, Tennessee, July 19, 1925. In this image, some of the outside observers are visible in the foreground. Photograph by Watson Davis. Courtesy of Smithsonian Institution Archives.

FIGURE 7.3. Worshippers assembled for a baptism in a stream near Dayton, Tennessee, July 19, 1925. Here, Preacher Joe Leffew (with light-colored hair and wearing a vest) is visible on the opposite bank standing at right. Photograph by Watson Davis. Courtesy of Smithsonian Institution Archives.

FIGURE 7.4. Worshippers observing a baptism near Dayton, Tennessee, July 19, 1925. This photograph may be the one that E. E. Slosson believed to represent "a woman seized by the power of the spirit and going into convulsions." Photograph by Watson Davis. Courtesy of Smithsonian Institution Archives.

FIGURE 7.5. Worshippers observing a baptism near Dayton, Tennessee, July 19, 1925. Preacher Joe Leffew is visible in the center, praying over the young woman seated on a rock. Photograph by Watson Davis. Courtesy of Smithsonian Institution Archives.

FIGURE 7.6. Worshippers observing a baptism near Dayton, Tennessee, July 19, 1925. Photograph by Watson Davis. Courtesy of Smithsonian Institution Archives.

Visible in several images is the preacher, just as described by Mencken and others—"an immensely tall and thin mountaineer in blue jeans, his collarless shirt open at the neck and his hair a tousled mop."[21] And then later the congregation witnesses the immersion and prayers from one side of the creek while Hays, Davis, and the other outsiders watch from the other.

FASCINATION

The baptism and nighttime services left strong impressions on those who observed them that July. Darrow even referred to the Holiness group in his appeal brief to the Tennessee Supreme Court ("You might as well say that the Commission should pass a law and adopt a geography which said that the world was round . . . or some of the 'Holy Rollers'—and I attended the 'Holy Rollers' on a former occasion here, and also at other places—who say the earth is flat.").[22] But they appear to have impressed even more someone who was not in Dayton—Watson's boss, E. E. Slosson.

Slosson was fascinated by the existence of such evangelical groups long before the 1925 trial, perhaps because their practices contrasted so sharply with the sobriety of his own Congregationalist services.[23] When Davis returned to Washington with the baptism photographs, Slosson began describing them excitedly to various correspondents.[24] He told G. L. Kleffer of the National Lutheran Council that because "most of their religious gymnastics take place in the churches or at night," Davis "was lucky to get some snapshots of a Holy Roller Baptism which shows one of their converts in convulsions."[25] He offered a set of the snapshots to Yale University psychologist Robert M. Yerkes "for psychological purposes": "One of them shows a woman seized by the power of the spirit and going into convulsions. Such pictures of the acrobatic school of Christianity are hard to get since their exercises are generally performed inside the church or at night."[26]

He also peddled them to potential publishers, explaining to W. L. Chenery, editor of *Collier's*, that "in this case we [that is, Davis] have snapped a woman just at the instant when she is 'seized by the power' so the picture is unique and has considerable anthropological value."[27] He later sent three images along with an essay to Chenery for the magazine's "Catching Up with the World" section, mentioning that one showed "the convulsive clinching of the hand and contraction of the arm characteristic of these seizures."[28]

What Slosson interpreted as the "seizures" of a worshipper may, in fact, have actually been a woman extending open-handed prayers over a newly or about-to-be baptized young girl, for Wacker has shown that "traditional social barriers" sometimes "crumbled" within the intensity of Holiness services. Women and children participated in testifying to their spiritual experiences and even occasionally preached to the congregants.[29]

To Slosson, the images possessed a seductive sensationalism. When he finally submitted the manuscript to *Collier's*, Slosson emphasized that he had prepared the essay with special care because "all such religious stuff carries dynamite and it is particularly difficult to write about such a formless and chaotic movement as that of the Holy Rollers without offending the more conservative fringe." Captioning might also pose a problem, he added, because "while 'Holy Rollers' is the common name, it is not acceptable to those to whom it is applied."[30] Nevertheless, he believed, it was important to pay attention to these expanding movements because of their rejection of modern science and technology.

INSIDERS' ROLE PLAY

That Sunday night, while a mimeograph machine rolled out copies of the scientists' statements, Darrow and the defense team held a secret role-playing session.

Darrow had earlier asked Charles Francis Potter to prepare a list of the Bible's "unscientific parts," that is, of passages and stories that contradicted modern scientific understanding, such as references to rabbits chewing cud.[31] Later that night, the defense team used those examples to prepare for a daring maneuver. They asked one of the scientists to play the role of Bryan, while they quizzed him for several hours on the contradictions between modern scientific knowledge and a literal interpretation of the Old Testament.[32]

It was important to have a script ready and to be well rehearsed. The defense was plotting to draw the real Bryan into the witness box.

CHAPTER 8 CONFRONTATION

Monday, July 20, 1925, was one of the hottest days of the summer. The prosecution's legal maneuvering dragged on through the morning. In the jammed courtroom, the heat was almost unbearable. Arthur Garfield Hays began to read the defense experts' statements into the record. After lunch, Judge Raulston announced that he was moving the proceedings outdoors.

What followed was one of the great moments in American legal history, not the least because of its setting. Prior to the trial, a temporary platform had been constructed in front of the courthouse. Charles Francis Potter had delivered his "Evolution" sermon from there the previous Sunday. William Jennings Bryan had preached from the stage, and even Judge Raulston himself. The platform offered a reasonable compromise to the court. The trial could continue in a cooler venue, and the judge could retain some dignity and authority elevated above the spectators.

Defense and prosecution teams were seated to the judge's right and left; a few policemen stood nearby to maintain order. The townspeople (and the jury) sat beneath the trees. "The spectators . . . instead of being only men," as had been usual inside the courthouse, the *New York Times* wrote, now

> were men, women and children, among them here or there a negro, sitting or standing on pine boards set on long saw-horses. In the rear of the crowd little children played on sea-saws [*sic*] made from the same boards and saw-horses. Small boys went through the crowd selling bottled pop. Most of the men wore hats and smoked.
>
> The seats were under the trees, but between them the sun blazed and scorched. Men and women hung from windows in the Court House.[1]

As Clarence Darrow recalled in his autobiography, "there were acres of audience, branching off into the surrounding streets, waiting for the curtain to rise."[2] In this extraordinary setting, framed by what Marcet

Haldeman-Julius called "a natural proscenium arch" of two great maple trees, the defense continued to read the experts' statements.[3]

In his classic history of the trial, Ray Ginger commented wryly that "constitutional law is a poor topic for a hot day."[4] Many journalists and photographers drifted back to the hotel's relative coolness. Then "a silence, broken only by the rustling of the maple trees, settled over the crowd" as Hays announced that the defense wanted to call Bryan to the stand.[5] Here, surely, would be the great confrontation, the battle everyone had come to see.

INTERROGATION

Over the objections of the rest of the prosecution team, Bryan agreed to testify. He walked over to the witness chair. Clarence Darrow rose and asked, "You have given considerable study to the Bible, haven't you, Mr. Bryan?" The examination that followed lasted about two hours.

Some parts of the trial and its proceedings had seemed like farce, like a haphazard jumble of politicians and lawyers jockeying for attention and press coverage. In this single moment, the action was reduced to two powerful antagonists, two men who had once fought together for liberal causes but were now enmeshed in a cultural struggle between fundamentalism and modernism. As the two-hour interrogation proceeded, the mood switched continually. Each man would grow impatient, then became calm. News accounts reported that they "glared at each other in undisguised displeasure."[6]

Most people in the modern world no longer believed that the earth was flat or encircled by the sun. How then, Darrow asked, could Bryan reconcile his literalist interpretation of the Bible with modern scientific knowledge about the solar system and with what intelligent, twentieth-century people accepted as fact about how the universe works? How could Bryan reconcile religion with modern science?

Bryan's response was simple: he saw no need to make such reconciliation.

> DARROW: Now, you say, the big fish swallowed Jonah, and he there remained how long? three days? and then he spewed him upon the land. You believe that the big fish was made to swallow Jonah?
>
> BRYAN: I am not prepared to say that; the Bible merely says it was done.

DARROW: You don't know whether it was the ordinary run of fish, or made for that purpose?

BRYAN: You may guess; you evolutionists guess.

Darrow continued to hammer Bryan with example after example of similar contradictions.

DARROW: The Bible says Joshua commanded the sun to stand still for the purpose of lengthening the day, doesn't it? and you believe it?

BRYAN: I do.

DARROW: Do you believe at that time the entire sun went around the earth?

BRYAN: No, I believe the earth goes around the sun.

DARROW: Do you believe that men who wrote it thought that the day could be lengthened or the sun stopped?

BRYAN: I don't know what they thought.[7]

At last, the moment that the drugstore conspirators had dreamed about had arrived. Here was the confrontation between giants, the clash upon the plains of justice, an instant of high drama for the press. And yet few photographs captured that encounter. Most surviving images record the action from above, as if the photographer had leaned out from a courthouse window, or they were taken from the back of the crowd. Watson Davis, however, had been standing for weeks on the side of the defense. Proximity finally paid off.

The first of the four photographs he took is fuzzy, as if the camera wiggled slightly. To look down into the viewfinder, Davis would have had to press that model of the Ica Victrix camera against his belly. He may have stood on a chair or ladder. Fortunately, by the third shot, he had steadied himself and sharpened the focus.[8]

Bryan's face tilts up expectantly for the next question. Darrow's powerful shoulders hunch slightly although his posture is neither wilted nor weak. Marcet Haldeman-Julius had said that Darrow's "whole torso" participated in every gesture he made in the courtroom. The defense attorney's head thrusts forward like a raptor reaching toward its prey. And Bryan seems transfixed.

The moment is simultaneously informal and, in retrospect, monumental. Through these images, we can almost feel the heat and sense the crowd's restlessness. As we peer more closely, we are transported to that

FIGURE 8.1. Clarence Darrow's interrogation of William Jennings Bryan at the Scopes trial, Dayton, Tennessee, July 20, 1925. As the temperature rose in the courtroom, Judge John R. Raulston decided to move the proceedings outside to a temporary platform erected next to the Rhea County Courthouse. Defense attorney Clarence Darrow (standing at right) began to interrogate prosecution team leader William Jennings Bryan (seated at left), and the crowd grew silent. Defense lawyers for Scopes (John R. Neal, Arthur Garfield Hays, and Dudley Field Malone) are visible seated to the extreme right in this and Figures 8.2 and 8.3. Behind the court reporters at the table, with his back to the camera, foreground left, is the defendant, John Thomas Scopes. Photograph by Watson Davis. Courtesy of Smithsonian Institution Archives.

FIGURE 8.2. Clarence Darrow's interrogation of William Jennings Bryan at the Scopes trial, Dayton, Tennessee, July 20, 1925. Photograph by Watson Davis. Courtesy of Smithsonian Institution Archives.

FIGURE 8.3. Clarence Darrow's interrogation of William Jennings Bryan at the Scopes trial, Dayton, Tennessee, July 20, 1925. Photograph by Watson Davis. Courtesy of Smithsonian Institution Archives.

sweltering afternoon. A straw hat hangs on a pole behind Bryan. Beyond the crowd at the left, at the edge of the courthouse yard, are the temporary privies with their banners exhorting all to believe.[9] One of the prosecutors holds his head in his hand. The defense attorneys seated at the right—Arthur Garfield Hays, Dudley Field Malone, and John Neal—lean forward. At a table near the foreground the court stenographers bend to transcribe the proceedings for posterity. And in the background we can see the crowd, the audience that Dayton had once hoped to transform into economic revival.

And where is Scopes in this extraordinary scene? He is seated just in front of the camera. He has joined the other observers along the sidelines. But in each of the photographs he took, Davis kept the back of Scopes's head in focus, as if reminding us not to forget the young man who was on trial.

CONVICTION

That night, Frank Thone and Watson Davis typed out their observations of what they had witnessed:

> Bryan's pitiful exhibition of ignorance under the skillful examination by Darrow was quite the most important evidence for the defense. . . . This dramatic episode . . . far outweighed the 60,000 words of scientific evidence that scientists and bible scholars had prepared in written statement.
>
> . . . Darrow's questioning made Bryan confess that he had little real knowledge of religion and science. Bryan admitted that he had never studied comparative religions . . . admitted never having studied philology . . . he declared he had never consulted any other authority than the Bible on the age of the earth or of mankind.[10]

Why did Bryan accept Darrow's challenge? Some historians suggest that Bryan's arrogance played a role in making him assume he could not lose; others cite Bryan's ambition to reclaim political power. The press attention to the trial might help to jumpstart another presidential campaign. Perhaps the temptation to engage with an old adversary was just irresistible. Whatever the motivation, there "was something pathetically humorous in Bryan's easy, almost gleeful acquiescence to [Darrow's] request," Marcet Haldeman-Julius wrote; "Even so has many an unsuspecting child climbed into the dentist's chair to descend from it later sadder and wiser."[11]

At the end of the examination, the judge adjourned the court. People young and old surged forward to shake Darrow's hand, looking "as though they were celebrating the victory of a championship team."[12] As the crowd left the courthouse lawn, some people were silent, some had "puzzled looks in their eyes—but many . . . left laughing."[13]

The local prosecutors conferred overnight and they decided, over Bryan's objections, to end the trial on Tuesday. The next morning, the jury deliberated for nine minutes and declared Scopes to be guilty.[14] The judge was about to pronounce sentence when Scopes, silent throughout the rest of the trial, asked if he could say a few words:

> Your honor, I feel that I have been convicted of violating an unjust statute. I will continue in the future, as I have in the past, to oppose this law in any way I can. Any other action would be in violation of my ideal of academic freedom—that is, to teach the truth as guaranteed in our constitution, of personal and religious freedom.[15]

Scopes was fined $100, which was paid by the *Baltimore Sun*. That night, somebody (some sources say the reporters, others credit the defense team) hired the village hall, brought in a band from Chattanooga, and gave the town a party.[16] Darrow danced.

BIDDING FAREWELL

On Wednesday, Dayton's merchants started to take down the welcome signs. Nellie Kenyon, the young Tennessee reporter who had first broken the story in May, saw Scopes "down town bright and early . . . bidding farewell to reporters, scientists and lawyers, not the slightest to be the 'convicted criminal.'"[17] Davis and Thone left town that Wednesday morning. Thanks to a clerk who routed them through Cincinnati, they did not reach Union Station until Thursday afternoon, July 23, but they, like all the Dayton visitors, had stories to tell and had formed strong opinions about the trial's significance. On Thursday evening, the National Women's Party held a reception in the garden of their Washington headquarters to honor their vice president, Doris Stevens, on her way back to New York. In her remarks to the gathering, Stevens called Judge Raulston "a sweet, lovable, incompetent old man who would not dare give a decision without recourse to prayer and Mr. Bryan's opinion," but she also praised the trial as "an unprecedented adventure in courtroom democracy."[18]

Despite the optimism of Scopes's supporters and despite Bryan's embarrassing performance, the long war for science and freedom of expression in the classroom was not over. That week, a group of antievolutionists initiated an effort to prohibit Smithsonian Institution research into human origins, and other politicians and citizens set out to make the teaching of evolution within District of Columbia schools an act of "disrespect for the Bible."[19] Bryan announced a national campaign to pass a constitutional amendment to prohibit the teaching of evolution.[20]

Later that week, Scopes and some of his Dayton buddies gathered the thousands of letters and business offers he had received since his arrest, piled them in the backyard of his rooming house, laughed at various marriage proposals, and tossed them all onto a bonfire.[21] It was a youthful gesture, one that signaled Scopes's desire to reclaim his life and leave celebrity behind.

BRYAN HEADS HOME

Dayton's dream of a quick economic boom had evaporated. The stage was being dismantled and the community was attempting to return to being "a sun-baked, slumberous, rather agreeable little country town among charming, wooded hills . . . a million miles away from anything urban, sophisticated, or exciting."[22] But that vision, too, proved illusory.

Bryan and his wife remained in the Rogerses' home for a few more days. He had been asked to deliver sermons to admirers in nearby towns and also wanted to complete the text of his response to Darrow. After traveling more than 200 miles on Saturday, July 25, Bryan attended church in Dayton on Sunday morning, ate a hearty lunch, lay down for an afternoon nap, and died in his sleep.[23]

The death of the "Great Commoner," and the plans for a train cortege to an elaborate Washington, D.C., funeral and burial in Arlington National Cemetery, preoccupied press and public for days. The residents of Dayton, shocked at the tragic ending to their summer fun, filed solemnly through the Rogerses' parlor to pay tribute to their fallen hero, and the American Legion arranged honor guards for the body.[24] Scopes came to pay his respects.[25] And when Bryan's coffin was being shifted onto the train by Dayton's Legionnaires, there helping to lift was everyone's friend, Post Commander George Washington Rappleyea.[26]

Bryan's death so soon after the end of the trial may have been one reason that the Davis photographs of the outdoor session were not published at the time. The loss of a man beloved by the masses and admired even by

FIGURE 8.4. William Jennings Bryan's coffin being loaded onto train, Dayton, Tennessee, July 1925. William Jennings Bryan had asked to be buried in Arlington National Cemetery, a honor to which he was entitled because he had served as U.S. secretary of state. His widow decided to take his body to Washington, D.C., via a special train, allowing supporters to pay their final respects along the way. This photograph shows the honor guard assembled by the Dayton American Legion Post. Visible in the center of the group (in uniform and riding boots) lifting the coffin is George Washington Rappleyea, one of the prime instigators of the trial that had brought Bryan to Dayton that summer. Courtesy of Special Collections Library, University of Tennessee, Knoxville.

his political foes shocked the nation. Some Bryan supporters even blamed his demise on the strain of that final interrogation.[27] Davis, Thone, and Slosson discussed the "Holy Rollers" photographs with many people, and Thone printed postcard versions of the "privies" photograph to share with friends, but there is no evidence that they attempted to promote or use the other images. Only a cryptic message, scribbled on a crumbling envelope and found eighty years later at the bottom of a dusty storage box ("Evolution Trial Open-Air Court Dayton, Tenn. July 1925 Note Bryan-Darrow etc. W.D."), hinted at the extraordinary moment in history captured in the darkened negatives.

CHAPTER 9 HEADING OUT OF TOWN

Years later, Watson Davis argued that the trial had provided "a great opportunity for those of scientific and liberal mind and inclination to speak to the public, through the press" and thereby draw attention to contemporary scientific thought, much as had the nineteenth-century debates between Thomas Henry Huxley and Bishop Samuel Wilberforce.[1] Such lofty comparisons aside, the events at Dayton had definitely given the two young reporters and their organization a chance to prove their mettle.

The presence of Davis and Frank Thone, as semiofficial representatives of the scientific establishment, helped lend credibility and authenticity to the defense. Their involvement allowed prominent scientists to participate in the trial at arm's length, without becoming muddied by the sensationalism. The journalists also enhanced their own professional reputations. Briton Hadden at *Time* told Davis that "we have been following your stories [from Dayton] in the press, by the way, with a great deal of interest, and have been making indirect use of them in our own columns—a practice to which I hope you will offer no objection."[2] And the attention attracted by the fledgling organization had pleased its benefactor. Midway through the trial, E. W. Scripps's personal assistant had wired praise from onboard the millionaire's yacht: "From the papers over this way, Science Service seems to be 'ringing the bell' on that Scopes trial."[3]

Perhaps most surprisingly, Science Service had made money, thereby reinforcing the idea that reliable and accurate science reporting could sell to the masses. Total income was $667.50 ($525.00 in special trial coverage and another $142.50 in individual stories); total expenses, including reimbursement to the ACLU for lodging in Defense Mansion, were $417.[4] In addition, favorable publicity to Science Service products dramatically expanded the client list and boosted the July income overall to 30 percent more than the previous year.

The greatest contribution made by Davis and Thone happened well out of the public eye, however. And it sprang from a simple, altruistic gesture.

PROTECTING SCOPES

As soon as the trial was over, "John Scopes" effectively split in two. The mythic Scopes, the hero or villain whose name and image still resonate in present-day debates over evolution, transcended into a popular culture character. That Scopes became the subject of cartoons, theater productions, novels, children's books, films, and countless retrospective and nostalgic articles about the summer of 1925. Meanwhile, the young man intent on education attempted to slip from public view, aided by friends who respected his decision and who worked to help him achieve his goals. The events in the years immediately following the trial, in fact, shed light on the enigmatic man caught up in a political and journalistic storm.

Most of the trial's organizers cared little about the long-term effect on the young teacher. Scopes had offered a convenient (and willing) excuse for political action, a role played by countless other accidental heroes past and present. Bryan told one of his supporters:

> It matters little, of course, what is done with Scopes. His fine will be paid by the New York society that is defending him and he will be a hero among those who think as he does. The trial will be a success in proportion as it enables the public to understand the two sides and the reasons on both sides.[5]

Even after the trial, an ACLU official admitted to neglecting the defendant, telling Scopes in late September:

> Once in a while when we think about the Tennessee evolution case we remember that there is a defendant. The advantage in having a defendant like ours is that he pops up in our consciousness only at the times when we need him or when our personal interest in him gives us the urge.[6]

Fortunately, others were not so callous.

Upon returning home, Davis and Thone, along with scientists Kirtley Mather and Maynard Metcalf, set about establishing a scholarship fund for Scopes.[7] The teacher was now out of a job. Employment anywhere in the South would be difficult. This "modest, unassuming young man" had voluntarily become the defendant "with no thought of personal reward, but with high courage in the face of public opinion," and he should not have to suffer financially because of participation in the case.[8] He had also consistently refused to capitalize on his fame, passing up many opportunities "to go a barnstorming, a la Red Grange, and . . . kept his head throughout the whole business," Thone wrote, so they "organized an informal committee to give him a bit of a boost" and "passed the hat."[9]

Davis served as secretary of the fund and Thone as treasurer. The two journalists appear to have developed and disseminated the solicitation letter, kept the campaign going, and taken responsibility for keeping in touch with Scopes. With the semester beginning soon, they needed to raise quickly a significant proportion of the $5,000 goal, so that Scopes could make his plans and enroll in a suitable graduate program. In Dayton, he had expressed a desire to study geology at the University of Chicago.

A group of nineteen regional chairmen was enlisted to tap potential donors. They included scientists who had volunteered to testify at Dayton (Winterton C. Curtis, William M. Goldsmith, William A. Kepner, Jacob G. Lipman, Kirtley M. Mather, Maynard M. Metcalf, and Wilbur A. Nelson) and other academic scientists such as David Starr Jordan at Stanford University. Thone also attempted to reach out to the press for help, suggesting that perhaps the *New York World* might assist in soliciting funds from New Yorkers because that paper had "maintained a very friendly and favorable attitude throughout the course of the Scopes trial."[10] In his letters soliciting contributions, Mather described eloquently how "unspoiled" Scopes was by the publicity and how the young man had refused "offers of glittering financial success if he would appear upon lecture platforms or vaudeville stage."[11]

Even though the argument for the fund was compelling, it was not easy to raise money. The ACLU, for example, apparently regarded the scholarship fund as competition for its campaign for reimbursement for the July trial and for the Tennessee Evolution Case Defense Fund it was establishing to support the appeal; John R. Neal even argued that any scholarship campaign should be delayed until after the appeal process was concluded, something that could take years.[12] Timing was also a problem, for the first newspaper announcements of the scholarship fund had unfortunately appeared the same day as major coverage of Bryan's death and funeral plans.[13]

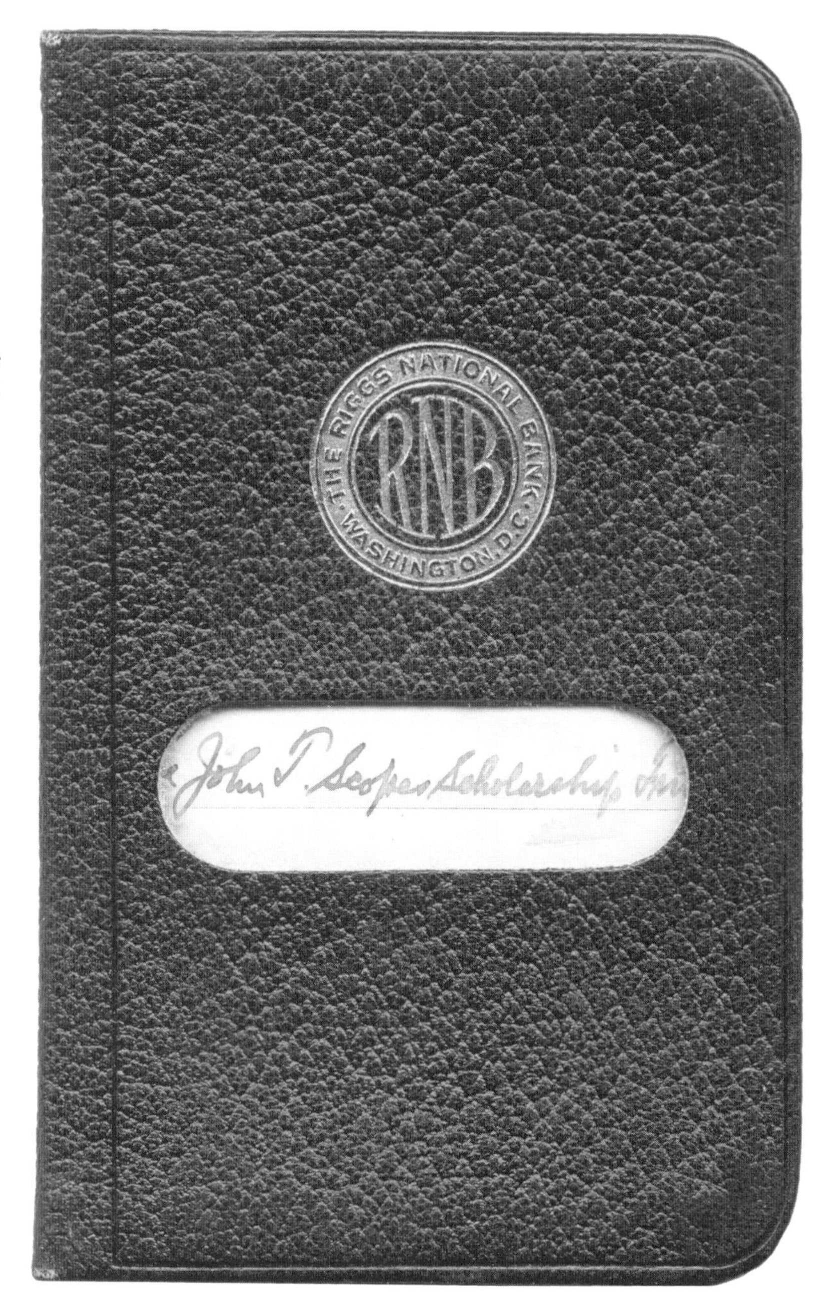

FIGURE 9.1.
Outside cover of bank book for John T. Scopes Scholarship Fund. Watson Davis served as secretary and Frank Thone as treasurer of the scholarship fund set up to enable John Thomas Scopes to attend graduate school at the University of Chicago. The account at Riggs National Bank in Washington, D.C., was managed by Thone. Courtesy of Smithsonian Institution Archives.

THE RIGGS NATIONAL BANK
WASHINGTON, D. C.

IN ACCOUNT WITH

1925

July	30	Dep	30
Aug	14	[illegible]	103
	14	[illegible]	1138 50
Sep	1	[illegible]	193 50
	11	[illegible]	67
	22	[illegible]	175
Oct.	5	c	308 50
	17	[illegible]	45 50
	28	[illegible]	57 50
Nov.	17	c	100
	28	[illegible]	100
Dec	19	c	42 50

Jan	13/26	c	68 —
Feb.	6	c	10 —
	13	c	10
Mar	11	c	100
July	16	c	2

FIGURE 9.2. Ledger pages from bank book for John T. Scopes Scholarship Fund. Courtesy of Smithsonian Institution Archives.

By mid-October, the John T. Scopes Scholarship Fund had accumulated $2,000 (half of which had been raised by David Starr Jordan), and the sponsors eventually raised only about $2,550.[14] Some of the most generous donations came from "persons outside the field of science," that is, from people who were, Thone acknowledged, "not under any immediate threat of loss of position or of legal penalty from laws similar to the Tennessee statute."[15] As Thone later explained to the editor of the *New Republic*, "We didn't get all the money needed for the three-year graduate course, for professors are poor men mostly."[16] Scopes would have to live frugally, supplement the fund with extra jobs, and, with luck, win a competitive fellowship for his final year of study.

MOVING ON

In mid-August 1925, Davis left on his first trip to Europe, sailing to England (to attend the British Association for the Advancement of Science meeting), then traveling to Berlin (for a brief audience with Albert Einstein), Geneva (to see the head of the League of Nations International Committee on Intellectual Cooperation), and Paris (where he had arranged to meet Marie Curie). By then, George Rappleyea had signed with the Adams Lecture Bureau to give talks on "The Battle of the Century" and "When Truth Is a Crime" for $100 each. Life in Dayton had finally returned to its pretrial languor. When Precious Rappleyea wrote to Davis to thank him for copies of photographs, she added: "Dayton is dead. That is the only way it can be described. But I will never forget the Evolution trial, will you?"[17]

In September, Scopes enrolled in graduate school at the University of Chicago. Once again, a Science Service stringer provided a key connection. Donald Glassman, a second-year student in the geology department, had contacted Science Service in mid-July 1925 to inquire about the type of articles they wanted. On the basis of Glassman's first submissions, Davis and Thone had invited him to send more.[18] Despite efforts by Darrow's office (which had issued a statement on Scopes's behalf, responding to news reports that he would be in Chicago), Thone knew that reporters would continue to hound the new student.[19] So, in September, when Thone accepted Glassman's most recent manuscripts, he added, "You may be interested to know that Mr. John T. Scopes of anti-evolution trial fame expects to take up the study of geology as a graduate student in Chicago this fall" and asked a favor:

> Please do what you can to protect him from the importunities of Chicago reporters, who may make an effort to get more copy out of him. He is a modest and unassuming young chap, and has been subjected to a great deal more of the limelight than he likes. I shall appreciate whatever courtesies and friendship you may show him.[20]

Glassman assured Thone that he would look out for Scopes.

In fact, the two graduate students became office mates in Rosenwald Hall. Glassman told Thone: "We had a four hour talk together the day I met him. He told me all about Darrow, Bryan, and the famous newspapermen who gathered at Dayton. He let U. and U. and P. and A. [two photo news services] get a picture of him but warned them never to show their scalps any more. I like him."[21]

From then on, Glassman's letters to Science Service almost always included a greeting from Scopes (including an occasional scribbled postscript). Thone, acknowledging submissions or sending payment, would send his best wishes in return. When Davis visited Chicago later that fall, he met with the two men, who "thoroughly enjoyed his company," even though Glassman received some tough criticism on his writing style and "even tho it did rain pitchforks all the time we were together."[22]

They appear to have been typical, high-spirited students. Scopes, who needed Thone's approval for any expenses charged to the scholarship fund, added a note to Thone at the bottom of a Glassman letter: "Hope to hell you are not having as much rain in Washington as we are here. No! Liquor is not cheap here. I am not going to tell a soul what I spent $12 for foolishly. JTS."[23] To which Thone responded at the end of his next letter to Glassman: "Give Scopes my best regards, and tell him I'll never ask about that 'foolishness' debit again."[24]

Sometimes the postscripts were friendly notes of encouragement—"Best personal regard to yourself and John Scopes. Tell him to cheer up: winter's nearly over."[25] Scopes, in fact, weathered the winter, did well in his classes, took a summer job with the Illinois State Geological Survey, and landed an instructorship for the next academic year. Things seemed to be looking up.

THE ANTIEVOLUTION MOVEMENT PERSISTS

That January, after the American Association for the Advancement of Science meeting in Kansas City, Watson Davis delivered a sermon on

"Science versus Religion" at L. M. Birkhead's church.[26] The scientific meeting, he explained to the congregants, had been "a great revival service" with the message that "ye shall know the truth and the truth shall make you free." There "is no true conflict between true science and true religion," he assured the congregation, "just as there can be no conflict between any facts of science."[27]

Others did not agree. The antievolution movement had not died with Bryan. As Davis preached understanding in Kansas City, the Supreme Kingdom, an organization founded by the Ku Klux Klan, was declaring that, in addition to its national campaign to remove textbooks that teach evolution from all schools and college, it would take its antievolution fight to the air. It was establishing a new broadcasting station to "carry to every nook and corner of the world the fight against the Darwinian theory."[28] Some fundamentalist lecturers were also demanding the exclusion of all biology from the school curriculum. Membership in the World's Christian Fundamentals Association, the group taking the lead in much of this effort, reportedly numbered around 6 million.[29] In California, Mississippi, Florida, Missouri, and South Carolina, efforts were under way to pass antievolution legislation. And in Arkansas, although the legislature had rejected one proposed resolution to prohibit teaching of "the theory or doctrine that mankind ascended or descended from a lower order of animals" in any state-funded institution, a group of antievolutionists, confident of support by church leaders, were vowing to carry the fight to the voters.[30]

MONEY MATTERS

One thing that kept Scopes in the public eye over the next year was the appeal, which Darrow and the ACLU team filed with the Tennessee Supreme Court in early January 1926. When that court handed down a decision on January 15, 1927, upholding the constitutionality of the law and overturning the conviction on a technicality, Scopes gave a single statement to the press and, with the assistance of fellow students, attempted to turn back to his studies.[31] As he wrote to Thone a few weeks later:

> The reporters also tried to keep me busy but I had too many good liars as friends. I was in Chicago all the time but every reporter that tried to get in touch with me learned that I was out of town. Just the other night I answered the door bell at the g.a. [Gamma Alpha] house & found that a reporter was very much interested to see me. I told him that Scopes

would not come down to see him so he left. I hated to do it, but I am tired of fooling with them.

Give my regards to Davis,
Your friend, J. T. Scopes[32]

After quoting that letter to Arthur Garfield Hays, Thone remarked: "That boy is wasted as a geologist, he ought to be working for the State Department."[33]

Diplomacy helps. But money matters.

Scopes's monthly room and board bill was $60; tuition was $66; clothing had to be replaced ($15 for shoes and hat, and $13 for underwear and socks); and young men need to go home occasionally (Christmas break trip, $40).[34] He told Kirtley Mather in April 1927 that he thought he had "enough money in the bank for another month unless some unexpected expense comes up," but "I had rather not run on too close a margin."[35] Scopes had been having trouble with his old eyeglasses, for example (tests and new glasses, $37). Although he still wanted to take a geology field course that summer, he was considering accepting a job with an oil company for a few years. That way, he could earn enough to finish at Chicago and perhaps even enroll for additional course work elsewhere. After all that had happened, Scopes remained a pragmatist: "There are so many things that a fellow would like to do, but cannot that it does not pay to plan too strong on ones [*sic*] likes and dislikes." Despite a heavy course load and the distractions of the appeal, he had done well in school, so he also applied for a fellowship to allow him to complete a third year.[36]

Unfortunately, once again, association with the "trial of the century" twisted the outcome. Although Scopes won the fellowship, when the president of the company donating the money heard that news, he called the university president and declared that no funds would go to any "damned evolutionist."[37] Scopes decided to leave school, but he had one more test to pass, this time related to weaknesses in the U.S. banking system.

In mid-April, Thone mailed Scopes a check for $538, the balance of the fund, to pay outstanding bills. Scopes would then deposit the remainder in a savings account (eventually about $200) to use when he returned to school. Once again, Thone apologized for not being able to raise enough money ("But as they carved on the cowboy's tombstone: 'we don our damdest; angels can't do no more.'") and sent wishes that the "Easter rabbits [*sic*] brings you plenty of gaudy eggs."[38] Scopes deposited the check. Then

HOTEL PENNSYLVANIA
NEW YORK

June 20

Dear Frank:

Arrived here this morning and I was surely glad to park myself where there ~~was~~ is a good bed and bath.

I certainly enjoyed seeing you and Paris again, and meeting your friend Mr. Stokley. I also enjoyed seeing Dr Lipmann (is that the way his name is spelled). In fact I had a wonderful day.

the bank informed him that the check, in transit back to Washington, had been on board a plane that crashed, and they demanded a duplicate check before crediting his account. Scopes had to hire a lawyer to write an affidavit to attest to the facts, authorize Thone to stop payment on the original, and hold harmless all parties should the original check ever reach the Washington bank. Thone offered to "advance you a little personally until the matter is settled," but Scopes declined, saying he was being allowed to run out his room and board bill, and other friends had been helping out.[39] Then Thone's Washington bank demanded that they post a bond of $1,000 "as insurance against the possibility of its having survived the fire, afterwards having been stolen, and some day being cashed." So, Scopes had to spend another $20 to post a bond, which Thone offered to advance: "It annoys me exceedingly that this unfortunate accident should have happened at the end of my administration of your fund, but that's the sort of thing we can expect in this somewhat messy cosmos. A few years hence we'll laugh over it, but for now I'm cussing."[40] Finally on May 31, Thone sent a duplicate of the long lost draft, astonished "at the clumsiness of our banking system when a small emergency like the present one arises."[41]

At that point, beleaguered by reporters, attacked by antievolutionists, and roiled by incompetent banks, Scopes was undoubtedly relieved to be leaving the country to take a two-year job as a Gulf Oil Company geologist in Venezuela.[42]

Throughout the past two years, he had been befriended by two men he had come to admire, men who had asked nothing of him and had actively sought to protect him from the trial's aftermath. His ship to South America sailed from New York on June 20. On the way north from his parents' home in Kentucky, Scopes took time to stop in Washington, to visit with Davis and Thone and meet the rest of the Science Service staff. From the Hotel Pennsylvania in New York, he wrote to thank them all for "a wonderful day."[43]

In the coming years, reporters would occasionally write to Davis or Thone seeking information on Scopes, and they continued to protect him. The man whose name and face had become known to most households in the country walked away from the "ballyhoo" age. As a *New York Times* writer told Thone, "It is refreshing to know that there is one person who dislikes publicity."[44]

FIGURE 9.3. Letter from John Thomas Scopes to Frank Thone, June 20, 1927, page 1. Scopes left the United States in June 1927 to work as an oil geologist for the Gulf Oil Company. On his way to New York City, where he boarded a merchant ship to sail to Venezuela, he stopped to visit Watson Davis and Frank Thone in Washington, D.C. He wrote this letter from his hotel to thank the journalists. The text reads: "Dear Frank: Arrived here this morning and I was surely glad to park myself where there is a good bed and bath. I certainly enjoyed seeing you and Davis again, and meeting your friend Mr. Stokley [James Stokley, an astronomer who had recently joined the Science Service staff]. I also enjoyed seeing Dr. Lipmann (is that the way his name is spelled) [Jacob G. Lipman, one of the expert witnesses in Dayton]. In fact I had a wonderful day." Courtesy of Smithsonian Institution Archives.

FIGURE 9.4.
Letter from John Thomas Scopes to Frank Thone, June 20, 1927, page 2. On the second page, Scopes dealt with some of the lingering financial details of the scholarship fund that Frank Thone had managed, saying that he was sending duplicates of the expense list to both Thone and Kirtley Mather, the Harvard geologist who served as president of the fund. Scopes then went over the details of the accounting process and asked that Thone notify him of the final amount to be placed in a savings account for use when he resumed his graduate work at the University of Chicago. He added his address in Venezuela on page three and, in fact, corresponded occasionally with Thone and Watson Davis during subsequent years. Courtesy of Smithsonian Institution Archives.

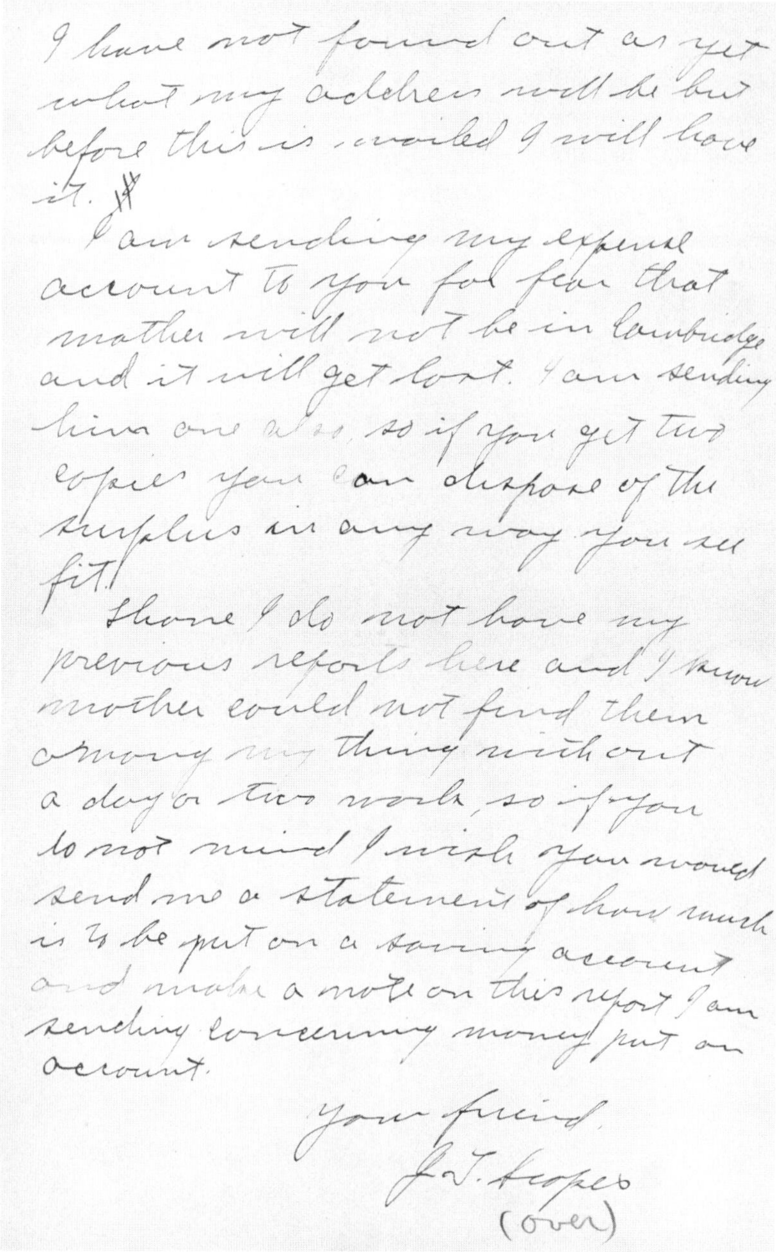

I have not found out as yet what my address will be but before this is mailed I will have it. ¶

I am sending my expense account to you for fear that mather will not be in Cambridge and it will get lost. I am sending him one also, so if you get two copies you can dispose of the surplus in any way you see fit.

Thone I do not have my previous reports here and I know mother could not find them among my thing without a day or two work, so if you do not mind I wish you would send me a statement of how much is to be put on a saving account and make a note on this report I am sending concerning money put on account.

Your friend.

J. T. Scopes

(over)

THE NEXT SCOPES

In 1928, the Arkansas Antievolution League successfully persuaded voters to pass an initiated act that made it illegal to teach evolution in any college or public school and forbade selection of textbooks that discussed the theory.[45] The ACLU immediately took steps to challenge that law "on behalf of some teacher who will maintain that his right to teach science is impaired by the operation of an unconstitutional statute."[46] When they eventually located a likely volunteer, the young man demanded the guarantee of "a fellowship, or a position as assistant in some college" afterwards, or what Thone derisively called an "asbestos martyr's robe."[47] Thone, of course, could not help contrasting the demand with "the conduct of young Scopes, who did not ask any questions about another job when he walked up to let the witch-burners have a whack at him."[48]

Thone's esteem and respect for Scopes remained undiminished to the end of his life. Thone had gotten to know the "real" Scopes. There was simply no comparison.

CHAPTER 10 LAST AND FIRST ACTS

"Drama, picture, story—America is all of these," E. Haldeman-Julius declared in the late 1920s: "As the biggest picture show on earth, it offers even greater variety: it is melodrama, pageant, sheer farce, comedy high and low. Its life each day exhibits all imaginable theatrical aspects."[1] The stock market crash of 1929 and the subsequent economic crisis took some of the comedy out of the pageant and, for many people, intensified the sense of melodrama.

In September 1930, when John Thomas Scopes returned to the University of Chicago to finish his graduate studies in geology, he was a married man. Although he completed all the courses and started the following summer on fieldwork, money continued to be a factor.[2] Frank Thone, Watson Davis, and others worked informally behind the scenes to find Scopes a teaching job, but the combination of history and circumstance increased the difficulty. One scientist noted that Scopes's "name coupled with that of Tennessee history barred him out repeatedly."[3]

The economic depression deepened. The young couple eventually went to live with Scopes's family in Kentucky. During the summer of 1932, Scopes willingly made one more foray into the public eye (something ignored even in his autobiography). He ran unsuccessfully for congressman-at-large from Kentucky on the Socialist ticket. This was the same national election that swept Franklin D. Roosevelt and many other Democrats into office throughout the country, but Scopes made a respectable showing and led all the other minority party candidates in the race.[4]

Scopes then returned to the oil business, working for companies in Houston and Louisiana, living quietly with his wife, Mildred, and two children, and even occasionally testifying before Congress as an oil expert, until his retirement in 1964. Although Rappleyea and Potter continued to exploit their association with the trial, Scopes avoided the spotlight for the next three decades, refusing to be interviewed and issuing only terse "not interested" statements when the state of Tennessee voted in 1935 to retain the antievolution law. He told one reporter in 1950 that "my friends who know about [the evolution trial] never bring it up in my presence."[5]

By the mid-1950s, few of the original participants survived. Attorneys Clarence Darrow, Arthur Garfield Hays, and Dudley Field Malone were gone; Frank Thone had suffered a fatal heart attack in 1949; and Charles Francis Potter had just officiated at L. M. Birkhead's funeral. Every few years, Watson Davis would patiently answer letters from students or historians inquiring about the trial, but his memories, too, were fading.[6]

Then, in 1955, a new play opened on Broadway and revived public interest in the case. Although the playwrights emphasized that the drama was not intended to be a historically accurate account of the Dayton trial—the script altered essential details and introduced new fictional characters—*Inherit the Wind* added to the developing mythology.[7]

Davis and Scopes appeared together on one more stage before they died. In 1960, in connection with the film version of *Inherit the Wind*, directed by Stanley Kramer, the Hollywood studio persuaded the Dayton Chamber of Commerce to engage once again in a publicity stunt—to host a "reunion" and local premiere of the movie.[8] The studio offered to fly Davis to Dayton, all expenses paid.[9]

At sixty-four, Davis was one of few original participants able to attend. Although the druggist F. R. Rogers was still filling prescriptions in town, only three of the original jurors remained. Scopes, apparently concerned about the mounting attacks on freedom of expression and various legislative proposals to prohibit the teaching of evolution, finally broke his silence. He stepped back into the spotlight, received the key to the city, and participated in the celebration of John T. Scopes Day.

On Thursday night, July 21, 1960, *Inherit the Wind* premiered at the Dayton Drive-in Theater. Scopes and Davis were entertained at receptions and made remarks from the stage. Photographs of the events show two aging, balding, and smiling men, enjoying the party.

Davis's cryptic notes from July 1960 are shaky and hard to read, but it is clear that his admiration for Scopes had not faded with time. Scopes, he scribbled, was still the same man who had acted with dignity and grace and had willingly made a "quiet gentle sacrifice of livelihood for principle."[10]

One sentence on the page of notes is sharp and unambiguous, however, and the strength of the statement is reflective of both the times and of Davis's own convictions: "If evolution teaches anything it is brotherhood of man."[11] E. E. Slosson, ever the optimist, had argued during the summer of 1925 that "it really does not matter much what becomes of evolution as a theory[,] whether people believe it or not, so long as they recognize evolution as a force and a fact and utilize it for the development of new and improved species of plants and animals, if not for the improvement of man himself."[12] Thirty-five years later, humankind still had a long way to go, but Davis knew how the first act must begin: to fight to preserve free expression and the open exchange of ideas, in and out of the classroom.

NOTES

ABBREVIATIONS

ACLU Archives (microfilm): American Civil Liberties Union Archives: The Roger Baldwin Years, 1917–1950 (microfilm), Manuscript Division, Library of Congress

Bryan Papers: Papers of William Jennings Bryan, 1877–1940, Manuscript Division, Library of Congress

Burbank Papers: Papers of Luther Burbank, 1830–1989, Manuscript Division, Library of Congress

Cattell Papers: James McKeen Cattell Papers, Manuscript Division, Library of Congress

Darrow Papers: Clarence Seward Darrow Papers, Manuscript Division, Library of Congress

Gruenberg Papers: Benjamin C. and Sidonie Matsner Gruenberg Papers, Manuscript Division, Library of Congress

Merriam Papers: John C. Merriam Papers, Manuscript Division, Library of Congress

Science Service Records (RU7091): Science Service Records, Record Unit 7091, Smithsonian Institution Archives

Science Service Records (97-020): Science Service Records, 1925–1966, Accession 97-020, Smithsonian Institution Archives

PREFACE

1 E. Haldeman-Julius, *Lessons Life Has Taught Me: Glimpses at the Fascinating Circus of Clowns and Philosophers* (Girard, Kans.: Haldeman-Julius Publications, 1928), 21.

2 Alan Trachtenberg, *Reading American Photographs: Images as History, Mathew Brady to Walker Evans* (New York: Hill and Wang, 1989), xiv.

CHAPTER 1. OPENING LINES

1 My summary of events during the spring and summer of 1925 relies on agreement among such sources as Ray Ginger, *Six Days or Forever? Tennessee vs. John Thomas Scopes* (New York: Oxford University Press, 1958); Norman Grebstein, ed., *Monkey Trial: The State of Tennessee vs. John Thomas Scopes* (Boston: Houghton Mifflin, 1960); Michael Kazin, *A Godly Hero: The Life of William Jennings Bryan* (New

York: Alfred A. Knopf, 2006); Edward Larson, *Summer for the Gods: The Scopes Trial and America's Continuing Debate over Science and Religion* (Cambridge, Mass.: Harvard University Press, 1997); Michael Lienesch, *In the Beginning: The Scopes Trial and the Making of the Antievolution Movement* (Chapel Hill: University of North Carolina Press, 2007); Jeffrey P. Moran, *The Scopes Trial: A Brief History with Documents* (New York: Palgrave, 2002); and John T. Scopes with James Presley, *Center of the Storm: Memoirs of John T. Scopes* (New York: Holt, Rinehart and Winston, 1967). See also more specialized works listed in the "Sources" section.

2 Vernon Kellogg to John C. Merriam, May 29, 1922, National Academies Archives, Central Policy Files, 1919–1923, Executive Board, Projects: Organic Evolution, 1922–1923. See also E. E. Slosson to Albert Barrows, May 3, 1922, Merriam Papers, Box 164, Folder "E. E. Slosson."

3 Kazin, *A Godly Hero*, 264. See also Lawrence W. Levine, *Defender of the Faith—William Jennings Bryan: The Last Decade, 1915–1925* (New York: Oxford University Press, 1965).

4 Kazin, *A Godly Hero*, esp. 263–264 and 275. For discussion of Bryan's concern about eugenics, see Moran, *The Scopes Trial*, 13–19.

5 Moran, *The Scopes Trial*; Kazin, *A Godly Hero*, esp. 262–263.

6 E. E. Slosson to John C. Merriam, December 14, 1921, Merriam Papers, Box 164, Folder "E. E. Slosson."

7 See, for example, Stephen Taber to Vernon Kellogg, March 16, 1922, National Academies Archives, Central Policy Files, 1919–1923, Executive Board, Projects: Organic Evolution.

8 E. E. Slosson to Albert Barrows, May 3, 1922, Merriam Papers, Box 164, Folder "E. E. Slosson." See also E. E. Slosson to Benjamin C. Gruenberg, May 12, 1923, Science Service Records (RU7091), Box 17, Folder 6.

9 See *AAAS Resolutions, 1920–25* (Washington, D.C.: American Association for the Advancement of Science, 1925), 66–67.

10 E. E. Slosson to John C. Merriam, December 14, 1921, Merriam Papers, Box 164, Folder "E. E. Slosson."

11 "Scientists, Publicists and Religious Leaders Declare That Science and Religion Are Allies, Not Enemies," *Daily Science News Bulletin*, no. 1134, mailed May 26, 1923, copy in Science Service Records (RU7091), Box 52, Folder 14.

12 See, for example, Robert A. Millikan to Charles D. Walcott, February 13, 1923, and Robert A. Millikan to John C. Merriam, February 13, 1923, enclosing draft statement, Merriam Papers, Box 125, Folder "Robert A. Millikan."

13 H. L. Mencken to Benjamin C. Gruenberg, May 27, 1924, Gruenberg Papers, Box 113.

14 See, for example, examples cited in Benjamin C. Gruenberg to E. E. Slosson, June 11, 1924, Gruenberg Papers, Box 113, Folder "General Correspondence 1924."

15 Benjamin C. Gruenberg to E. E. Slosson, September 27, 1924, Gruenberg Papers, Box 113, Folder "Slossen [*sic*], E. E."

16 Charles A. Israel, *Before Scopes: Evangelicalism, Education, and Evolution in Tennessee, 1870–1925* (Athens: University of Georgia Press, 2004), esp. 128–155.

17 Jeannette Keith, *Country People in the New South: Tennessee's Upper Cumberland* (Chapel Hill: University of North Carolina Press, 1995), 197.

18 Ibid., 199.

19 The Tennessee secretary of state's office mailed a typed copy of House Bill 185 to Watson Davis in March 1925. On the bottom of a National Research Council routing slip attached to the typescript, returning it to Watson Davis, was a comment by scientist Vernon Kellogg: "Weird stuff!" Science Service Records (RU7091), Box 78, Folder 7.

20 Kenneth K. Bailey, "The Enactment of Tennessee's Antievolution Law," *Journal of Southern History* 16 (November 1950): 472–490. See Keith, *Country People*, 203–210, for a profile of J. W. Butler and his motivations in introducing the legislation.

21 Levine, *Defender of the Faith*, 328.

22 Robinson was described as "a one-man chamber of commerce." Ginger, *Six Days or Forever?* 354.

23 "F. E. Robinson Co./The Hustling Druggist/Main St. Near Post Office/Dayton, Tennessee" (advertisement), *Dayton Herald*, July 23, 1925.

24 The Tennessee Textbook Commission had adopted George William Hunter's *Civic Biology* as an official high school biology text in 1919. As in many small towns, Robinson's store was the official depository for purchase of schoolbooks.

25 Per Scopes's own account of events in Scopes, *Center of the Storm*.

26 George Washington Rappleyea to Forrest Bailey, August 7, 1925, p. 1, ACLU Archives (microfilm), Reel 38, Volume 274.

27 Forrest Bailey to the Committee on Academic Freedom, May 27, 1925, p. 1, ACLU Archives (microfilm), Reel 38, Volume 273.

28 They initially had difficulty locating Bryan. On May 14, Sue K. Hicks wrote to Bryan: "We have been trying to get in touch with you by wire to ask you to become associated with us in the prosecution of the case of the State against J. T. Scopes, charged with violation of the anti-evolution law, but our wires did not reach you." Sue K. Hicks to William Jennings Bryan, in Miami, Florida, May 14, 1925, Bryan Papers, Box 47, Folder 1. By then, Bryan had already been quoted in the paper as saying he would be involved in the trial.

CHAPTER 2. EDUCATION, PERSUASION, AND PASSION

1 Clipping in Science Service Records (RU7091), Box 44, Folder 2.

2 For a more complete history of the establishment of Science Service, see Marcel

Chotkowski LaFollette, *Science on the Air: Popularizers and Personalities on Radio and Early Television* (Chicago: University of Chicago Press, 2008). For discussion of the organization's approach to "selling" science news, see Marcel C. LaFollette, "Taking Science to the Marketplace: Examples of Science Service's Presentation of Chemistry during the 1930s," *HYLE* 12 (2006): 67–97.

3 E. E. Slosson to Thomas T. Coke, February 7, 1921, Science Service Records (RU7091), Box 7, Folder 1.

4 Scripps gave an annual donation of $30,000 from 1921 until his death in 1926. His will provided a similar annual endowment until 1956.

5 Davis was, for example, an active member of the Penguins, a "National Liberal Club," located in downtown Washington, D.C. See club brochure in Science Service Records (RU7091), Box 81, Folder 1.

6 Slosson died in 1929; Davis was eventually named director in 1933 and remained in that job until 1966.

7 Statement by J. Speed Rogers, Frank Thone, John Gray, and T. H. Hubbel, January 1924, Merriam Papers, Box 172, Folder "Thone, Frank."

8 E. E. Slosson to secretary, American Civil Liberties Union, May 15, 1925, Science Service Records (RU7091), Box 44, Folder 2.

9 Olin Templin to E. E. Slosson, November 14, 1921, Science Service Records (RU7091), Box 13, Folder 1.

10 E. E. Slosson to Emily Green Balch, March 27, 1924, Science Service Records (RU7091), Box 21, Folder 11.

11 Many years later, responding to a question about Slosson's religious affiliation, Frank Thone said that as "a deacon in the Mount Pleasant Congregationalist Church, [Slosson] occasionally took over the pulpit when the minister was absent. I can give you first-hand 'ear-witness' testimony that his sermons were excellent." Frank Thone to Earnest W. Lundeen, June 1, 1945, Science Service Records (RU7091), Box 270, Folder 4.

12 "Emmerich Lecture Bureau, Inc., presents Edwin E. Slosson" (brochure), Science Service Records (RU7091), Box 78, Folder 5.

13 E. E. Slosson to Benjamin C. Gruenberg, May 12, 1923, Science Service Records (RU7091), Box 17, Folder 6.

14 "Scientists, Publicists, and Religious Leaders Declare That Science and Religion Are Allies, Not Enemies," *Daily Science News Bulletin*, no. 1134, mailed May 26, 1923, Science Service Records (RU7091), Box 52, Folder 14.

15 See also "The American Association for the Advancement of Science and the Legislation against the Teaching of Evolution," *Science* 61 (May 29, 1925), supplement section, x. On May 25, an Associated Press story stated that Rappleyea "also quoted Dr. Watson Davis, editor of Science Service of Washington, as saying 'we are coming to your support 14,300 strong.'" Clippings in Science Service Records (RU7091), Box 44, Folder 3. George Washington Rappleyea used

similar language in a telegram to Science Service on May 25, reporting on the grand jury session. "Scientists Pledge Support to Tennessee Professor Arrested for Teaching Evolution," *Science News-Letter* 6 (June 6, 1925): 1–2.

16 *Daily Science News Bulletin*, mailed May 20, 1925, Science Service Records (RU7091), Box 78, Folder 7.

17 Watson Davis to John Martin, October 18, 1929, Science Service Records (RU7091), Box 108, Folder 6. After the trial, E. E. Slosson also refused to serve on the ACLU advisory committee for the Scopes appeal. Watson Davis telegram to ACLU, July 27, 1925, ACLU Archives (microfilm), Reel 38, Volume 274.

18 E. W. Scripps to E. E. Slosson, August 1, 1921, Science Service Records (RU7091), Box 12, Folder 2.

19 See descriptions in Watson Davis to William E. Ritter, January 7, 1925, and Joseph Barnett to Watson Davis, February 18, 1925, Science Service Records (RU7091), Box 82, Folder 6. See also "Navy Dirigible Eclipse Expedition Saw Brilliant Corona by Watson Davis, Managing Editor, Science Service" and "Press Representative on Navy Dirigible USS Los Angeles during Eclipse Flight," *Science News-Letter* 6 (February 14, 1925): 4–5; "Scientists on Los Angeles Praise First Dirigible Eclipse Flight," *New York Times*, January 25, 1925.

20 Richard Streckfuss, "Objectivity in Journalism: A Search and a Reassessment," *Journalism Quarterly* 67 (Winter 1990): 973–983.

21 By June 23, at least four major newspapers (*New York Herald-Tribune*, *Brooklyn Eagle*, *Atlanta Constitution*, and *Chicago Daily News*) had signed on for Science Service dispatches from the trial.

22 Science Service telegram to ACLU, May 21, 1925, Science Service Records (RU7091), Box 44, Folder 3.

23 ACLU to telegram to E. E. Slosson, May 22, 1925, and Watson Davis telegram to G. W. Rappleyea, May 22, 1925, Science Service Records (RU7091), Box 44, Folder 3.

24 Watson Davis telegram to George W. Rappleyea, May 22, 1925, Science Service Records (RU7091), Box 44, Folder 3.

25 George W. Rappleyea telegram to Watson Davis, May 22, 1925, Science Service Records (RU7091), Box 44, Folder 3.

26 Watson Davis to George W. Rappleyea, May 22, 1925, Science Service Records (RU7091), Box 44, Folder 3.

27 Watson Davis to members of the Executive Committee, May 23, 1925, Science Service Records (RU7091), Box 2, Folder 1.

28 Mark Sullivan to Watson Davis, May 25, 1925, Science Service Records (RU7091), Box 81, Folder 8.

29 E. E. Slosson to Mark Sullivan, May 27, 1925, Science Service Records (RU7091), Box 81, Folder 8. See also E. E. Slosson to Watson Davis, May 27, 1925, Science Service Records (RU7091), Box 103, Folder 2.

30 E. E. Slosson to Watson Davis, May 27, 1925, Science Service Records (RU7091), Box 103, Folder 2.

31 "Minutes of Executive Committee Meeting of Science Service," June 2, 1925, Science Service Records (RU7091), Box 2, Folder 1. John C. Merriam, who apparently wrote the resolution, told Watson Davis that the minutes were incorrect and should have read not "for expenses and full reporting" but "for expenses *in* full reporting." John C. Merriam to Watson Davis, June 15, 1925, Merriam Papers, Box 51, Folder "Davis, Watson, 1922–1930."

CHAPTER 3. DETOUR TO DAYTON

1 Watson Davis to Edna Jackson, May 25, 1925, Science Service Records (RU7091), Box 80, Folder 2. Making such a detour would not have been easy in 1925 and would probably have required Davis to change trains several times and make a special roundtrip to Dayton by rail or car from either Cincinnati or Chattanooga. I thank railroad historian John H. White for this information.

2 Watson Davis to Frank Thone, June 1, 1925, Science Service Records (RU7091), Box 82, Folder 3. Thone was in Iowa, visiting his parents.

3 Davis ordered the equipment in March 1925 from Philadelphia dealer W. J. McCann on the recommendation of Science Service staff member and astronomer James Stokley. Total cost for the camera, lens, nine plate holders, soft leather wallet case, developing tank, and film pack adapter was $34.10. Watson Davis to W. J. McCann, March 26, 1925, Science Service Records (RU7091), Box 90, Folder 6.

4 H. L. Mencken, "Mencken Finds Daytonians Full of Sickening Doubts about Value of Publicity," originally published in the *Baltimore Evening Sun*, July 9, 1925, reprinted in H. L. Mencken, *A Religious Orgy in Tennessee: A Reporter's Account of the Scopes Monkey Trial* (Hoboken, N.J.: Melville House Publishing, 2006), 29.

5 E. Haldeman-Julius, *Lessons Life Has Taught Me: Glimpses at the Fascinating Circus of Clowns and Philosophers* (Girard, Kans.: Haldeman-Julius Publications, 1928), 21.

6 William Jennings Bryan had delivered the commencement address at Scopes's high school in Marion, Illinois. "Scopes Here, Shyly Defends Evolution," *New York Times*, June 7, 1925.

7 John Thomas Scopes, "Reflections—Forty Years After," in Jerry R. Tompkins, ed., *D-Days at Dayton* (Baton Rouge: Louisiana State University Press, 1965), 18. See also John T. Scopes, with James Presley, *Center of the Storm: Memoirs of John T. Scopes* (New York: Holt, Rinehart and Winston, 1967).

8 See, for example, the Associated Press interview in "Scopes Upholds Evolution but Denies Monkey Theory," *Atlanta Constitution*, May 28, 1925.

9 Watson Davis, "Scientific Aspects of the Scopes Trial" (typescript), March 31, 1964, Science Service Records (97-020), Box 1, Folder 4.

10 "Scopes Here, Shyly Defends Evolution."
11 Jack Lait, "Debate Evolution after Trial Over," *Chattanooga News*, July 13, 1925.
12 See, for example, Kirtley F. Mather to James McKeen Cattell, July 28, 1925, Cattell Papers, Box 135, Folder "Mather, Kirtley."
13 Draft for Chapter 29 ("Dayton") of *The Story of My Life*, Darrow Papers, Container 9, Folder 8.
14 Philip Kinsley, "Indict Scopes for Teaching Evolution Law," *Chicago Daily Tribune*, May 16, 1925; and Philip Kinsley, "Lines Drawn for Epochal Battle over Darwin," *Los Angeles Times*, May 26, 1925. See also Philip Kinsley, "Famous Test Suit of Evolution Had Trivial Origin," *Washington Post*, May 31, 1925.
15 Arthur Garfield Hays, *Let Freedom Ring* (New York: Boni and Liveright, 1937), 33.
16 Alfred W. McCann to William Jennings Bryan, June 30, 1925, Bryan Papers, Box 47, Folder 3.
17 Raymond Clapper, "Scopes Would Devote Life to Study of Evolution," *Atlanta Constitution*, July 23, 1925.
18 The lecture bureau that handled E. E. Slosson even tried (unsuccessfully) to enlist Watson Davis in convincing Scopes to sign on for a tour after the trial. F. J. Emmerich to Watson Davis, July 7, 1925, Science Service Records (RU7091), Box 78, Folder 5.
19 "Tennessee Jury Hears Evolution Charges To-day," *New York Herald Tribune*, May 25, 1925.
20 George Britt, "Just Thought He Would Get Fired," *Chattanooga News*, June 10, 1925. Britt, a journalist for the N.E.A. Service owned by Scripps-Howard, interviewed Scopes in New York.
21 Ibid.
22 Roderick Nash, *The Nervous Generation: American Thought, 1917–1930* (New York: Rand McNally, 1970), 5 and 127. See also Joshua Gamson, *Claims to Fame: Celebrity in Contemporary America* (Berkeley: University of California Press, 1994); and Richard Schickel, *Intimate Strangers: The Culture of Celebrity* (Garden City, N.Y.: Doubleday, 1985).
23 "Scopes Here, Shyly Defends Evolution."
24 Marcet Haldeman-Julius, "Impressions of the Scopes Trial," *Haldeman-Julius Monthly* 2 (September 1925): 325.
25 Rappleyea spent much of the 1930s as a successful powerboat salesman and marina manager, and he was an extremely skilled sailor. Dayton appears to be one of the few inland places he ever lived. See, for example, "The Editor an Early Bird," *Eureka News Bulletin* 1 (October 1942): 7; and "Early Days of Aviation," *Eureka News Bulletin* 2 (February 1943): 25–28.
26 L. Sprague De Camp, who interviewed and corresponded with Rappleyea from around 1958 to 1965, says that Rappleyea "worked his way through an irregular

college education." L. Sprague De Camp, *The Great Monkey Trial* (Garden City, N.Y.: Doubleday, 1968), 4.

27 Other Legion members at the time were A. W. Morgan and druggist Fred R. Rogers.

28 "Geologic Section, Oil and Gas Well of Dryden Development Co., Chattanooga Tenn., Showing Possible Location of Petroleum, Geo W Rappleyea, Engineer & Geologist," 1922, Science Service Records (RU7091), Box 44, Folder 4.

29 "Setting the Stage for a Big Battle between Science and Religion," *Kansas City Star*, June 7, 1925.

30 Kinsley, "Famous Test Suit of Evolution Had Trivial Origin."

31 H.L. Mencken, "The Scopes Trial: Tennessee in the Frying Pan," originally published in the *Baltimore Evening Sun*, July 20, 1925, reprinted in Mencken, *A Religious Orgy in Tennessee*, 100–101.

32 Haldeman-Julius, "Impressions of the Scopes Trial," 325.

33 "Rappleyea's Razzberry," *Time*, June 1, 1925.

34 W. C. Morgan, "The Dayton Coal and Iron Company Limited," in Bettye J. Broyles, *History of Rhea County, Tennessee* (Dayton, Tenn.: Rhea County Historical and Genealogical Society, 1991), 103–110. See also T. J. Campbell, *Records of Rhea: A Condensed County History* (Dayton, Tenn.: Rhea Publishing, 1940), 64–83.

35 Salt's father, Sir Titus Salt, had made a fortune in textile manufacturing.

36 Morgan estimated the first expenditures to have been between $1.5 and $2 million (the equivalent of some tens of millions of dollars today). Morgan, "The Dayton Coal and Iron Company Limited," 104.

37 "Strawberry Industry," in Broyles, *History of Rhea County*, 103–110. See also Campbell, *Records of Rhea*, 99–101.

38 "J. Astor Squires Buys Coal Tract," *New York Times*, July 14, 1921; "Financial Notes," *New York Times*, August 23, 1922; "Rail Suit Compromised," *New York Times*, October 7, 1922.

39 Roland Marchand, *Advertising the American Dream: Making Way for Modernity, 1920–1940* (Berkeley: University of California Press, 1985), 1 (chapter title).

40 "Rappleyea's Razzberry."

41 All quotes from *Why Dayton—Of All Places?* (Dayton, Tenn.: privately printed, 1925); copy in Darrow Papers.

42 Ibid., 15.

43 "Knoxville Wants Evolution Case; Dayton Favored," *Daily (Chattanooga) Times*, May 19, 1925.

44 "Evolution Taught at Central High; Students Called before Grand Jury; Instructor Prosecuted in Test Case," *Daily (Chattanooga) Times*, May 19, 1925.

45 "Tennessee Cities Fighting to Stage Evolution Test Case," *Atlanta Constitution*, May 20, 1925.

46 See "Geologic Section, Oil and Gas Well of Dryden Development Co."

47 Watson Davis, "The Rocks and Hills of Dayton Testify for Evolution," *Science News-Letter* 6 (June 25, 1925): 2–3. See also *Science* 61 (June 19, 1925): xi.

48 Davis, "The Rocks and Hills of Dayton Testify for Evolution," 2.

49 Before leaving Los Angeles, Davis quickly wrote and sold stories about the earthquake. See, for example, "Willis Says Earth Moved 16 Inches," *New York Times*, July 1, 1925.

50 Such rhetorical strategies were being routinely employed in 1920s press accounts of celebrities. See Charles A. Ponce de Leon, *Self-Exposure: Human-Interest Journalism and the Emergence of Celebrity in America, 1890–1940* (Chapel Hill: University of North Carolina Press, 2002), 6.

51 Westbrook Pegler, "Scopes, Storm Center in Tennessee War on Evolution, Pays First Visit to New York," *Atlanta Constitution*, June 8, 1925.

52 "Messrs. Darrow, Colby and Malone Survey Chassis of Ancient Steeds," (Benton Harbor, Michigan) *News-Palladium*, June 12, 1925; and "Dr. Osborn Advises Scopes on Defense," *New York Times*, June 9, 1925.

53 "Scopes Rests Hope in U.S. Constitution and Supreme Court," *Washington Post*, June 13, 1925; and "Scopes Visits Capitol," *Washington Evening Star*, June 13, 1925.

54 In 1925, the Library of Congress's displays were in its main structure (completed in 1897 and now referred to as the Jefferson Building), the original copy of the U.S. Constitution was housed in the library, and the Supreme Court (its present building not yet constructed) was in a chamber of the U.S. Capitol.

CHAPTER 4. PARTICIPANTS AND OBSERVERS

1 Watson Davis to Charles Adams, July 30, 1925, Science Service Records (RU7091), Box 75, Folder 10; and news article draft ("It is the hope . . ."), Science Service Records (RU7091), Box 365, Folder 3.

2 News article draft ("It is the hope . . ."), Science Service Records (RU7091), Box 365, Folder 3.

3 The clients were the *New York Herald Tribune*, *Brooklyn Eagle*, *Buffalo Evening News*, *Atlanta Journal*, *Chattanooga Times*, *Chicago Daily News*, *Birmingham News*, *Kansas City Star*, and *Oklahoma City Times*.

4 Watson Davis to George Washington Rappleyea, and George Washington Rappleyea to Watson Davis, July 6, 1925, Science Service Records (RU7091), Box 81, Folder 4.

5 See notes by Watson Davis, Frank Thone, and others in Science Service Records (RU7091), Box 43, Folder 9.

6 List, in Watson Davis's handwriting, labeled "Dr. White," Science Service Records (RU7091), Box 43, Folder 9.

7 Watson Davis telegram to George Rappleyea, July 7, 1925, Science Service Records (RU7091), Box 43, Folder 10.

8 Telegrams in Science Service Records (RU7091), Box 44, Folder 1. Handwriting samples in the Darrow Papers show that this was not Darrow's signature, although it is possible that he authorized Science Service to act on his behalf in communicating with potential scientific witnesses.

9 See telegrams and correspondence in Science Service Records (RU7091), Box 44, Folder 1.

10 Watson Davis to Jerry R. Tompkins, January 29, 1963, Science Service Records (97–020), Box 1, Folder 4.

11 Davis later said that "most of the work was done on the train going from Washington to Chattanooga" with Hays and Malone and that he and Thone had compiled the lists of potential experts. Watson Davis to Jerry R. Tompkins, January 29, 1963, Science Service Records (97–020), Box 1, Folder 4.

12 Watson Davis to Manager, Western Union Telegraph Co., July 8, 1925, Science Service Records (RU7091), Box 82, Folder 6.

13 Typed telegram drafts, labeled to be charged to Cumberland Coal & Iron Company, and indicating they were sent from Dayton, were annotated in Watson Davis's handwriting.

14 Watson Davis, "Scientific Aspects of the Scopes Trial" (typescript), March 31, 1964, Science Service Records (97–020), Box 1, Folder 4.

15 F. R. Rogers to William Jennings Bryan, June 2, 1925, Bryan Papers, Box 47, Folder 1.

16 Bryan requested four rooms and a bath. William Jennings Bryan to Sue K. Hicks, June 7, 1925, and similar correspondence about arrangements, Bryan Papers, Box 47, Folder 1.

17 "Attorneys to Be Housed in Mansion on Hill," *Atlanta Constitution*, June 30, 1925.

18 Allene Sumner, "Ghosts Are Evicted to Make Room for Great Army of Scientists to Visit Dayton Court," *Mobile News-Item*, July 6, 1925.

19 Marcet Haldeman-Julius, "Impressions of the Scopes Trial," *Haldeman-Julius Monthly* 2 (September 1925): 325.

20 H. L. Mencken, "Law and Freedom, Mencken Discovers, Yield Place to Holy Writ in Rhea County," originally published in the *Baltimore Evening Sun*, July 15, 1925, reprinted in H. L. Mencken, *A Religious Orgy in Tennessee: A Reporter's Account of the Scopes Monkey Trial* (Hoboken, N.J.: Melville House Publishing, 2006), 73.

21 Haldeman-Julius, "Impressions of the Scopes Trial," 325.

22 Charles Francis Potter, *The Preacher and I: An Autobiography* (New York: Crown Publishers, 1951).

23 See Watson Davis to Arthur Garfield Hays, July 29, 1925, Science Service Records (RU7091), Box 79, Folder 4. Davis said that he relinquished his reser-

vation at the Hotel Aqua to a Unitarian minister who contracted typhoid and died not long after the trial ended. Davis, "Scientific Aspects of the Scopes Trial." The disease was a concern for all present in Dayton that summer. Journalist Raymond Clapper, for example, came down with typhoid after the trial but survived. "Clapper Works Weeks on Evolution Trial with Typhoid Fever," *UNIPEP* 2 (August 10, 1925): 1, 4.

24 Watson Davis, "Scientific Men and the Defense of Mr. Scopes," *Science* 62 (August 7, 1925): 130. See also *New York Times*, July 14, 1925.

25 Jeffrey P. Moran, *The Scopes Trial: A Brief History with Documents* (New York: Palgrave, 2002), 3–9.

26 Potter, *The Preacher and I*, 273.

27 See biographical information about Marcet (1887–1941) and Emanuel (1888–1951) Haldeman-Julius on the Kansas State Historical Society Web site, http://www.kshs.org. See also Gene DeGruson, "Anna Marcet and Emanuel Haldeman-Julius: An Afterword," on the Web site for the E. Haldeman-Julius Collection, Leonard H. Axe Library, Pittsburg State University, http://library.pittstate.edu/spcoll.

28 Haldeman-Julius, "Impressions of the Scopes Trial," 323.

29 "Man vs. Monkey Tilt Has Aspect of Circus," *Los Angeles Times*, July 5, 1925; and advertisement, *New York Times*, June 28, 1925.

30 Westbrook Pegler, "Excitement of Shelby Missing in Dayton on Eve of Trial," *Atlanta Constitution*, July 10, 1925. Early reports said that the turnout for opening day barely reached 500, which was certainly not the thousands the business leaders had envisioned. Philip Kinsley, "Battle for Evolution Opens in Court Today," *Los Angeles Times*, July 10, 1925. See also John T. Moutoux, "Concessionaires in Dayton Fail to Reap Profits," *Mobile News-Item*, July 11, 1925; and "Size of Crowd Fails to Meet Anticipations of Dayton Folk," *Atlanta Constitution*, July 15, 1925.

31 "Cranks and Freaks Flock to Dayton," *New York Times*, July 11, 1925.

32 Pegler reported that "most of the visitors are newspaper writers and photographers." Pegler, "Excitement of Shelby Missing in Dayton on Eve of Trial." About 150 members of the press had requested credentials as of the trial's opening day, but there is no record or official count of those who actually arrived.

33 "Statement of defense counsel for the press—proposed by Darrow and approved by other counsel," July 2, 1925, ACLU Archives (microfilm), Reel 38, Volume 274.

34 Clarence Darrow, *The Story of My Life* (New York: Charles Scribner's Sons, 1932), 262–263.

35 Arthur Garfield Hays, *Let Freedom Ring* (New York: Boni and Liveright, 1937), 35.

36 "2,000,000 Words Wired to the Press," *New York Times*, July 22, 1925.

37 Ibid.

38 Davis and his boss E. E. Slosson had actually been corresponding with Mencken for some years as the writer established *The American Mercury*.

39 S. L. Harrison, "The Scopes 'Monkey Trial' Revisited: Mencken and the Art of Edmund Duffy," *Journal of American Culture* 17 (1994): 55–63.

40 June Adamson, "Nellie Kenyon and the Scopes 'Monkey Trial,'" *Journalism History* 2 (Autumn 1975): 88–89, 97.

41 In 1920, Brisbane told an audience of University of Kansas journalism students that "if people don't have some kind of excitement they will have another, and I believe that a paper that gives legitimate excitement to people renders a public service. . . . The newspaper furnishes the vaudeville in the lives of a great many people." Quoted in John H. McCool, "Yellow Fellow," January 19, 1910, "This Week in Kansas History," http://www.kuhistory.com.

42 William E. Ritter to Watson Davis, October 13, 1927, Science Service Records (RU7091), Box 90, Folder 6.

43 The staff member ("MG"—Mary McGrath) was asked to examine "editorial attitude, news treatment, and general remarks" in July 1925 coverage of papers with 50,000–74,999 circulation (*Galveston News*, *Galveston Tribune*, *Wheeling Register*), 250,000–299,999 circulation (*Atlanta Journal*, *Atlanta Georgian*, *Toledo Blade*, *Toledo News-Bee*), 400,000–499,999 circulation (*New Orleans Times-Picayune*, *New Orleans State*), more than 1,000,000 circulation (*New York Journal*, *Cleveland Press*, *Cleveland Plain Dealer*, *Los Angeles Times*, *New York Times*, *New York Herald-Tribune*, *Chicago Tribune*, *Philadelphia Evening Bulletin*, *Baltimore Sun*), the major Tennessee papers (*Chattanooga News*, *Chattanooga Times*, *Johnson City Chronicle*, *Johnson City Staff-News*, *Kingsport Times*, *Knoxville Journal*, *Knoxville News*, *Knoxville Sentinel*, *Memphis Commercial Appeal*, *Memphis Press*, *Memphis Scimitar*, *Nashville Banner*, and *Nashville Tennessean*), and others added to broaden the geographic reach (*Seattle Post-Intelligencer*, *Pensacola Journal*, *Salt Lake City Tribune*). Davis's notes indicate that he personally did the analysis of the *Baltimore Sun* and *Chattanooga News*. Science Service Records (RU7091), Box 90, Folder 6.

44 Watson Davis, "Attitude of Reporters Covering Scopes Trial," 1927, Science Service Records (RU7091), Box 443, Folder 9.

45 Quin Ryan, WGN's announcer, boarded at the home of druggist F. E. Robinson and took his meals at the home of banker A. P. Haggard, whose son Wallace Haggard was one of the local prosecutors.

46 "W-G-N to Bring Evolution Case to Your Home," *Chicago Tribune*, June 28, 1925; "Broadcast of Scopes Trial Unprecedented," *Chicago Daily Tribune*, July 5, 1925; "W-G-N Will Take Scopes Trial to Record Audience," *Chicago Daily Tribune*, July 12, 1925; and Elmer Douglass, "Elmer Haunts Scopes' Trial Radio Waves," *Chicago Daily Tribune*, July 16, 1925. See also *WGN: A Pictorial History* (Chicago: WGN, 1961), 21–22; James Walter Wesolowski, "Before Canon 35: WGN Broadcasts

the Monkey Trial," *Journalism History* 2 (Autumn 1975): 76–79, 86–87; Erik Barnouw, *A Tower in Babel: A History of Broadcasting in the United States to 1933* (New York: Oxford University Press, 1966), 196–197; and Donald G. Godfrey and Frederic A. Leigh, eds., *Historical Dictionary of American Radio* (Westport, Conn.: Greenwood Press, 1998), 359.

47 "W-G-N Will Take Scopes Trial to Record Audience." WGN broadcasts stopped on Friday, July 17.

48 "Evolution Sidelights," *Atlanta Constitution*, July 12, 1925. The microphones were placed on top of 4-foot lengths of wooden four-by-fours. "Intimate Glimpses of Life as It Is Being Lived at Dayton, Tenn.," *Mobile News-Item*, July 12, 1925.

49 "Inside the Loud Speaker with Quin A. Ryan, the Voice of W-G-N," *Chicago Daily Tribune*, July 26, 1925. WGN strove "to give listeners everywhere the same advantage they would have if they were present at the trial," "Broadcast of Scopes Trial Unprecedented."

50 "Storm Silences W-G-N, but You'll Hear Trial Today," *Chicago Daily Tribune*, July 14, 1925.

51 Helen Miles Davis to Watson Davis, July 15, 1925, Science Service Records (RU7091), Box 103, Folder 2.

CHAPTER 5. RELIGIOUS FEELING

1 Frank R. Kent, "On the Dayton Firing Line," *New Republic*, July 29, 1925, 259.

2 Arthur Garfield Hays, *Let Freedom Ring* (New York: Boni and Liveright, 1937), 41 and 18.

3 H. L. Mencken, "Law and Freedom, Mencken Discovers, Yield Place to Holy Writ in Rhea County," originally published in the *Baltimore Evening Sun*, July 15, 1925, reprinted in H. L. Mencken, *A Religious Orgy in Tennessee: A Reporter's Account of the Scopes Monkey Trial* (Hoboken, N.J.: Melville House Publishing, 2006), 71.

4 Charles Francis Potter, *The Preacher and I: An Autobiography* (New York: Crown Publishers, 1951). See also Charles Francis Potter, *Humanism: A New Religion* (New York: Simon and Schuster, 1930), esp. 131–133.

5 See the biographical profile of Birkhead at http://www.uua.org.

6 Birkhead, who actively criticized many of the fundamentalist evangelists, provided help and inspiration to Sinclair Lewis as the novelist wrote *Elmer Gantry*. See Marcet Haldeman-Julius, "Sinclair Lewis and a Liberal Preacher," in Henry W. Thurston (with a Symposium on Churches by Various Writers), *Why I Did Not Enter the Methodist Ministry*, Little Blue Book no. 1217 (Girard, Kans.: Haldeman-Julius Publications, 1927), 29–58.

7 Ibid, 36.

8 Ibid.

9 Ibid., 37.

10 A. W. Ogden, "Pastor Where Scopes Attends Church Is Fundamentalist," *Knoxville Sentinel*, July 10, 1925.

11 "Scopes' Sponsor Is Relieved of Post in Sunday School," *Atlanta Constitution*, June 23, 1925; and "Scopes' Aide Quits as Head of Dayton S.S.," *Atlanta Constitution*, June 29, 1925.

12 "Dayton Minister Responsible for Scopes Evolution Test," *Atlanta Constitution*, July 2, 1925; *New York Times*, July 3, 1925; and "Pastor Involved in Scopes Case Painting Church," *Mobile News-Item*, July 7, 1925.

13 Associated Press, dateline Dayton, Tenn., *New York Times*, July 3, 1925.

14 "Trial Stirs Local Storm," *New York Times*, July 13, 1925.

15 "Evolution Row Rends Pastorate," *Los Angeles Times*, July 13, 1925; Philip Kinsley, "Church Quarrel Ushers in Scopes Trial Today," *Los Angeles Times*, July 13, 1925; "Ostracize Dayton Pastor," *Los Angeles Times*, July 13, 1925; and "Trial Stirs Local Storm."

16 "Ostracize Dayton Pastor." "Byrd Won't Quit Ministry," *New York Times*, July 20, 1925. Byrd left Dayton soon thereafter and was offered a special scholarship from the Methodist church but according to L. Sprague De Camp, he went instead to a church in Alabama. L. Sprague De Camp, *The Great Monkey Trial* (Garden City, N.Y.: Doubleday, 1968).

17 Kinsley, "Church Quarrel Ushers in Scopes Trial Today."

18 Ibid.

19 Potter, *The Preacher and I*, 285.

20 Ibid.

21 "Bryan Preaches Old-Time Creed on Court Lawn," *Chicago Daily Tribune*, July 13, 1925.

22 John Herrick, "Now an Anti-Evolution College," *Chicago Daily Tribune*, July 13, 1925. The Associated Press also quoted him as saying "It is b-y-Byrd. . . . I have evolved from b-i-Bird, which has feathers, you know." "Pastor of Dayton Church Quits When Congregation Bans Modernist Lecture," *Atlanta Constitution*, July 13, 1925.

CHAPTER 6. THE SCIENTISTS COME TO TOWN

1 News article draft ("All day long . . ."), written on or about July 15, 1925, Science Service Records (RU7091), Box 365, Folder 3.

2 Years later, Davis explained that they knew not to ask scientists who "were extremely busy" or that they "thought would not come to Dayton." Watson Davis to Jerry R. Tompkins, January 29, 1963, Science Service Records (97–020), Box 1, Folder 4.

3 Philip Kinsley, "Bryan Keen for Battle," *Los Angeles Times*, July 8, 1925. See also "Bryan in Dayton, Calls Scopes Trial Duel to Death," *New York Times*, July 8, 1925.

4 Watson Davis's handwritten notes for telegram to Science Service office, Science Service Records (RU7091), Box 43, Folder 9.

5 Norman Grebstein, ed., *Monkey Trial: The State of Tennessee vs. John Thomas Scopes* (Boston: Houghton Mifflin, 1960), 88.

6 There is a discrepancy between the published transcript and other evidence about who was actually sworn in that day for the defense and who was actually in Dayton and submitted testimony. The published transcript, for example, lists Kirtley Mather as sworn that Wednesday but he did not actually arrive in Dayton until Saturday morning. See Kirtley F. Mather, "Geology and Genesis," in Jerry R. Tompkins, ed., *D-Days at Dayton: Reflections on the Scopes Trial* (Baton Rouge: Louisiana State University Press, 1965), 88; and July 8 and July 14, 1925, telegrams from Kirtley Mather in Science Service Records (RU7091), Boxes 43 and 44. See also Mather's statements to L. Sprague De Camp, as quoted in *The Great Monkey Trial* (Garden City, N.Y.: Doubleday, 1968). A formal statement by Dean Shailer Mathews of the University of Chicago Divinity School appears in the transcript, but although he was willing to come, Mathews was not actually in Dayton for the trial.

7 News article draft ("Never before was . . ."), written July 15, 1925, Science Service Records (RU7091), Box 365, Folder 3.

8 Statements by Dean Shailer Mathews of the University of Chicago Divinity School and botanist Luther Burbank, neither of whom traveled to Dayton, were submitted for the record. See Burbank correspondence with Classic Publishing Company, Brunswick, Georgia, 1925, Burbank Papers, Box 4, Folder "C, Misc., 1917–1926."

9 "Experts Called to Scopes Trial," *Atlanta Constitution*, June 26, 1925. For Mather's views on science and religion, see Kirtley F. Mather, "Evolution and Religion," *Scientific Monthly* 21 (September 1925): 322–328.

10 *The World's Most Famous Court Trial: Tennessee Evolution Case*, 3rd ed. (Cincinnati, Ohio: National Book Co., 1925), 263.

11 Horatio Hackett Newman, *Evolution, Genetics, and Eugenics*, 3rd ed. (Chicago: University of Chicago Press, 1938), 46.

12 News article draft ("All day long . . .").

13 Winterton C. Curtis, *Science and Human Affairs from the Standpoint of Biology* (New York: Harcourt, Brace and Company, 1922).

14 W. C. Curtis to Dudley Field Malone (telegram), July 8, 1925, Science Service Records (RU7091), Box 44, Folder 1.

15 See brief discussion of Goldsmith in Gary R. Enz, "Religion in Kansas," *Kansas History* 28 (Summer 2005): 140.

16 William M. Goldsmith, *Evolution and Christianity* (Girard, Kans.: Haldeman-Julius Publications, 1924); and William M. Goldsmith, *Evolution or Christianity, God or Darwin?* (St. Louis, Mo.: The Anderson Press, 1924).

17 See telegrams in Science Service Records (RU7091), Boxes 43 and 44.

18 Jeffrey P. Moran, *The Scopes Trial: A Brief History with Documents* (New York: Palgrave, 2002), 12. See also Shailer Mathews, *New Faith for Old: An Autobiography* (New York: Macmillan, 1936), 226–229, and the biographical portrait of Mathews at http://www.lib.uchicago.edu/projects/centcat/ (accessed July 6, 2007).

19 News article draft ("Just what does this Dayton performance . . ."), July 17, 1925, Science Service Records (RU7091), Box 365, Folder 3.

20 Ibid.

21 Frank Thone to J. H. Aldrich, August 27, 1925, Science Service Records (RU7091), Box 75, Folder 10.

22 Ibid.

23 Maynard M. Metcalf to Michael I. Pupin, July 17, 1925, Darrow Papers, Container 5, Folder 1.

24 Frank Thone to J. H. Aldrich, August 27, 1925, Science Service Records (RU7091), Box 75, Folder 10.

25 Marcet Haldeman-Julius, "Impressions of the Scopes Trial," *Haldeman-Julius Monthly* 2 (September 1925): 330.

26 News article draft ("All day long . . .").

CHAPTER 7. SUNDAY EXCURSIONS

1 The leader of the group is named in several contemporary sources and, with no other evidence available, I have used that name and spelling. As with many independent religious sects and congregations, archival records are minimal. For an extensive study of such groups, see Grant Wacker, *Heaven Below: Early Pentecostals and American Culture* (Cambridge, Mass.: Harvard University Press, 2001).

2 Ibid. See also T. Rennie Warburton, "Holiness Religion: An Anomaly of Sectarian Typologies," *Journal for the Scientific Study of Religion* 8 (Spring 1969): 130–139.

3 News article draft ("Far from Dayton . . ."), written on or about July 13, 1925, Science Service Records (RU7091), Box 365, Folder 3. This draft, which described the activities and theological origins of the sect, was clearly written by Thone. A shorter, revised version ("The earth is flat . . .") was published with the byline "Watson Davis" and apparently drew formal condemnation from a group of Baptist ministers in Chattanooga. See draft of a letter to the editor of the *Chattanooga Times* bylined "Rev. Watson Davis," Science Service Records (RU7091), Box 365, Folder 3.

4 Wacker, *Heaven Below*, 1, 99–102.

5 News article draft ("The earth is flat . . ."), written on or about July 13, 1925, Science Service Records (RU7091), Box 365, Folder 3.

6 Ronald L. Numbers, "Creation, Evolution, and Holy Ghost Religion: Holiness and Pentecostal Responses to Darwinism," *Religion and American Culture* 2 (Summer 1992): 131.
7 News article draft ("The earth is flat . . .").
8 Arthur Garfield Hays, *Let Freedom Ring* (New York: Boni and Liveright, 1937), 39.
9 Allene M. Sumner, "The Holy Rollers on Shinbone Ridge," *Nation* (July 29, 1925), 137.
10 Clarence Darrow, *The Story of My Life* (New York: Charles Scribner's Sons, 1932), 261.
11 H. L. Mencken, "Yearning Mountaineers' Souls Need Reconversion Nightly, Mencken Finds," originally published in *Baltimore Evening Sun*, July 13, 1925, reprinted in H. L. Mencken, *A Religious Orgy in Tennessee: A Reporter's Account of the Scopes Monkey Trial* (Hoboken, N.J.: Melville House Publishing, 2006), 49–59.
12 "Delirium in Dayton," *Spectator*, July 18, 1925, 94–95.
13 Frank R. Kent, "On the Dayton Firing Line," *New Republic* (July 29, 1925), 259.
14 Thone is probably referring to *Pinus torreana*, which grew near San Diego (where Thone had worked) and which botanists then believed had once had a more extensive range.
15 News article draft ("Far from Dayton . . .").
16 "Bryan and Darrow Wage War of Words in Trial Interlude," *New York Times*, July 19, 1925.
17 Hays, *Let Freedom Ring*, 39.
18 "Bryan and Darrow Wage War of Words in Trial Interlude."
19 "Dayton Fortified against Last Shock of Darrow Attack," *New York Times*, July 20, 1925.
20 Hays, *Let Freedom Ring*, 40.
21 Mencken, "Yearning Mountaineers' Souls," 53.
22 Darrow's argument to the Supreme Court of the State of Tennessee, Nashville, June 1, 1926, in *John Thomas Scopes vs. State of Tennessee*, p. 13, Darrow Papers, Container 5, Folder 6.
23 "When I was in Texas and Arkansas, I was told that the Holy Rollers were making more rapid gains than any other religion in those states and in southern Indiana." E. E. Slosson to Emily Green Balch, February 19, 1924, Science Service Records (RU7091), Box 21, Folder 11.
24 Davis sent copies of the photographs to others who were apparently there that day, such as Arthur Garfield Hays, Wilson Midgley (the U.S. correspondent for the *London Daily News*), and E. Haldeman-Julius.
25 E. E. Slosson to G. L. Kleffer, August 14, 1925, Science Service Records (RU7091), Box 45, Folder 6.

26 E. E. Slosson to Robert M. Yerkes, August 17, 1925, Science Service Records (RU7091), Box 30, Folder 12.
27 E. E. Slosson to W. L. Chenery, August 7, 1925, Science Service Records (RU7091), Box 90, Folder 8.
28 E. E. Slosson to *Collier's* Editorial Department, August 21, 1925, Science Service Records (RU7091), Box 90, Folder 8.
29 Wacker, *Heaven Below*, 104.
30 E. E. Slosson to *Collier's* Editorial Department, August 21, 1925.
31 Charles Francis Potter, *The Preacher and I: An Autobiography* (New York: Crown Publishers, 1951), 275.
32 The surviving evidence is conflicting here on whether the scientist who participated in this role-play was Maynard Metcalf or Kirtley Mather.

CHAPTER 8. CONFRONTATION

1 "Big Crowd Watches Trial under Trees," *New York Times*, July 21, 1925. For other accounts of that afternoon session, see "Dramatic Scenes in Trial," *New York Times*, July 21, 1925; Philip Kinsley, "Bryan-Darrow Quiz a Duel," *Chicago Daily Tribune*, July 21, 1925; and "Bryan and Darrow in Bitter Religious Clash as Commoner Is Quizzed on His Bible Views," *Atlanta Constitution*, July 21, 1925.
2 Clarence Darrow, *The Story of My Life* (New York: Charles Scribner's Sons, 1932), 262.
3 Marcet Haldeman-Julius, "Impressions of the Scopes Trial," *Haldeman-Julius Monthly* 2 (September 1925): 345. According to Grebstein, the statements read outdoors that afternoon, along with the biographies of each expert, "constituted approximately 35,000 words." Norman Grebstein, ed., *Monkey Trial: The State of Tennessee vs. John Thomas Scopes* (Boston: Houghton Mifflin, 1960), 143.
4 Ray Ginger, *Six Days or Forever? Tennessee v. John Thomas Scopes* (New York: Oxford University Press, 1958).
5 "Bryan and Darrow in Bitter Religious Clash."
6 Ibid.
7 *The State of Tennessee vs. John Thomas Scopes*, transcript for July 20, 1925, as reprinted in Grebstein, *Monkey Trial*. See also the testimony for July 20, 1925, as reprinted in Clarence Darrow, *The Famous Examination of Bryan at the Scopes Evolution Trial*, Little Blue Book no. 1424 (Girard, Kans.: Haldeman-Julius Publications, 1929).
8 My thanks to Shannon Perich and David Haberstitch, associate curators of photography at the Smithsonian Institution, for assistance early in my research in interpreting the variations in this sequence and explaining the operation of the Ica Victrix camera.
9 This aspect of the image was revealed through computer enhancement per-

formed for the Smithsonian Institution Archives by Smithsonian Photographic Services.

10 News article draft ("Bryan's pitiful exhibition . . ."), written by Frank Thone, July 20, 1925, and published with Watson Davis's byline in the *Chattanooga Times*, July 21, 1925, Science Service Records (RU7091), Box 365, Folder 3.

11 Marcet Haldeman-Julius, "Impressions of the Scopes Trial," 314.

12 News article draft ("Bryan's pitiful exhibition . . .").

13 "Big Crowd Watches Trial under Trees."

14 Philip Kinsley, "Conviction Hurls Evolution into Larger Field," *Washington Post*, July 22, 1925.

15 Grebstein, *Monkey Trial*, 176.

16 Raymond Clapper, "Scopes Would Devote Life to Study of Evolution," *Atlanta Constitution*, July 23, 1925.

17 Ibid.; and Nellie Kenyon, "Dayton Is Glad Trial at End: Faith Unmoved," *Chattanooga News*, July 22, 1925.

18 "Miss Stevens, Scopes Aid, To Speak Here," *Washington* Post, July 22, 1925; "Capital Society Events," *Washington Post*, July 23, 1925; and "Miss Stevens Tells of Trial at Dayton," *Washington Post*, July 24, 1925.

19 "Scopes' Verdict Makes Evolution National Issue; Antis to Ask Congress to Declare for Bible Version," *Washington Herald*, July 22, 1925; Kinsley, "Conviction Hurls Evolution into Larger Field"; and "Fight on Teaching of Evolution Here Begun in Courts," *Washington Post*, July 23, 1925.

20 "New Evolution Case Defense Seeking to Avoid 'Forum' Here," *Washington Post*, July 24, 1925.

21 John T. Scopes with James Presley, *Center of the Storm: Memoirs of John T. Scopes* (New York: Holt, Rinehart and Winston, 1967), 191–195.

22 Michael Williams, "At Dayton, Tennessee," *Commonweal*, July 22, 1925, 262.

23 See George Washington Rappleyea telegram to Science Service, July 26, 1925, Science Service Records (RU7091), Box 81, Folder 4; and "William Jennings Bryan Dies of Apoplexy as He Is Taking Afternoon Nap at Dayton," *Atlanta Constitution*, July 27, 1925. Rappleyea was reportedly one of the first Dayton citizens to call at the Rogerses' home after Bryan's death. See *Los Angeles Evening Express*, July 17, 1925.

24 "Coffin in Little Cottage," *New York Times*, July 28, 1925; and "Mountain Folk Will View Commoner's Remains on Lawn Today," *Washington Post*, July 28, 1925.

25 "Scopes Views Body of Dead Prosecutor," *Chattanooga News*, July 29, 1925.

26 The honor guard also included Clay Green, Creed Parham, Ed Morgan, Will Fisher, Wilfred Ault, and storekeeper Gordon Darwin.

27 See, for example, "The Dayton Battle May Have Been Bryan's Doom," *Literary Digest*, August 8, 1925.

CHAPTER 9. HEADING OUT OF TOWN

1 Watson Davis, "Scientific Aspects of the Scopes Trial" (typescript), March 31, 1964, Science Service Records (97-020), Box 1, Folder 4. This is the draft that Davis sent to Jerry R. Tompkins for publication in *D-Days at Dayton: Reflections on the Scopes Trial* (Baton Rouge: Louisiana State University Press, 1965).

2 Briton Hadden to Watson Davis, July 17, 1925, and Watson Davis to Briton Hadden, July 25, 1925, Science Service Records (RU7091), Box 79, Folder 4. Davis told Hadden that he was "flattered."

3 H. L. Smithton to Watson Davis, July 17, 1925, Science Service Records (RU7091), Box 81, Folder 8.

4 "Comparative Statement, Science Service, Incorporated, for July 1, to July 31, 1925," Science Service Records (RU7091), Box 78, Folder 3.

5 William Jennings Bryan to Ed Howe (editor), June 30, 1925, Bryan Papers, Box 47, Folder 3.

6 Forrest Bailey to John T. Scopes, September 29, 1925, ACLU Archives (microfilm), Reel 38, Volume 274.

7 "The John T. Scopes Scholarship Fund," *Science*, July 31, 1925, 105; "Scientists Launch Drive for Scopes Scholarship," *Science News-Letter* 7 (August 8, 1925): 10; and Raymond Clapper, "Scopes Would Devote Life to Study of Evolution," *Atlanta Constitution*, July 23, 1925. Citing a 1958 interview with Kirtley Mather, L. Sprague De Camp stated that Davis initiated the scholarship campaign. See L. Sprague De Camp, *The Great Monkey Trial* (Garden City, N.Y.: Doubleday, 1968), 414. De Camp never mentioned Thone anywhere in his book (Thone had died over a decade earlier), however, and I believe it is far more likely, given the correspondence record, Thone's association with the University of Chicago, and Thone's many other charitable actions in later years, that the idea came from *both* Davis and Thone.

8 Solicitation letter for the John T. Scopes Scholarship Fund, 1925. See copies in Smithsonian Institution Archives, Records of the Office of the Secretary, Records, 1925–1949 (RU46), Box 30, Folder 1; and in Merriam Papers, Box 161, Folder 7.

9 Frank Thone to editor of *New Republic*, May 16, 1926, Science Service Records (RU7091), Box 88, Folder 1.

10 Frank Thone to James McKeen Cattell, August 3, 1925, Cattell Papers, Box 154, Folder 11.

11 See, for example, Kirtley Mather to James McKeen Cattell, July 28, 1925, Cattell Papers, Box 135, Folder "Mather, Kirtley."

12 John R. Neal to Forrest Bailey, August 14, 1925, and other correspondence in ACLU Archives (microfilm), Reel 38, Volume 274.

13 Notices appeared in the Washington, D.C., papers on July 27, for example. Also

see "As $5,000 to Give Scopes Science Course; Scientists Find Him Unharmed by Limelight," *New York Times*, August 8, 1925.

14 In the first months, contributions from nonscientists outnumbered those of scientists. See Frank Thone to Watson Davis, August 20, 1925, Science Service Records (RU7091), Box 78, Folder 3; Frank Thone to G. W. Rappleyea, August 20, 1925, Science Service Records (RU7091), Box 81, Folder 4; and Frank Thone to Watson Davis, September 1, 1925, Science Service Records (RU7091), Box 78, Folder 3. Both Thone and Mather attempted to raise money at scientific meetings during the winter of 1925–1926. See Frank Thone to Kirtley F. Mather, January 13, 1926, Denison University Archives, 12P M1 Box 19, K. B. Bork Biography of Kirtley Mather, "Scopes, John T./Scopes Trial." Amount of $2,550 calculated from deposits listed in the bank book for John T. Scopes Scholarship Fund, the Riggs National Bank, Washington, D.C. (Science Service Records [RU7091], Box 90, Folder 5). Most contributions were quite modest. Luther Burbank, for example, responded immediately with a check for $25. Burbank Papers, Box 10, Folder "T Misc 1891–1926."

15 "Scopes to Study Geology at Chicago University," *Science News-Letter* 7 (October 3, 1925): 9.

16 Thone was writing to correct a magazine's statement that Scopes had received a handsome fellowship from the National Research Council. Frank Thone to editor of *New Republic*, May 16, 1926, Science Service Records (RU7091), Box 88, Folder 1.

17 Precious Rappleyea to Watson Davis, August 10, 1925, Science Service Records (RU7091), Box 78, Folder 3.

18 See correspondence from and to Donald Glassman, Science Service Records (RU7091), Box 83, Folder 1.

19 "Scopes Seeks Degree at Midway School," *Chicago Daily Tribune*, September 26, 1925; and "Scopes to Study in Chicago," *New York Times*, September 26, 1925. In his autobiography, Scopes praised the warmth and support that the Darrows provided while in Chicago. Scopes roomed in a house owned by the secretary of Shailer Mathews.

20 Frank Thone to Donald Glassman, September 14, 1925, Science Service Records (RU7091), Box 83, Folder 1.

21 Donald Glassman to Frank Thone, September 28, 1925, and October 3, 1925, Science Service Records (RU7091), Box 83, Folder 1. The P and A photograph is probably the one that was published in the *Los Angeles Times*, October 7, 1925, showing a very solemn Scopes in a three-piece suit.

22 Donald Glassman to Frank Thone, November 9, 1925, Science Service Records (RU7091), Box 83, Folder 1; Watson Davis handwritten notes dated November 7, 1925; and Watson Davis to George W. Rappleyea, November 14, 1925, Sci-

ence Service Records (RU7091), Box 81, Folder 4. At the end of his second year, Glassman left graduate school and became a reporter for the *Cincinnati Post*. See his letters in Science Service Records (RU7091), Box 28, Folder 6.

23 Donald Glassman to Frank Thone, November 12, 1925, Science Service Records (RU7091), Box 83, Folder 1. The "foolishness" charge appears to have been a running joke and was characteristic of Scopes's cordial relationship with Thone. Per other correspondence, the twelve dollars had not been spent on liquor. J. T. Scopes to Kirtley F. Mather, November 30, 1925, and Frank Thone to Kirtley F. Mather, December 7, 1925, Denison University Archives, 12P M1 Box 19, K. B. Bork Biography of Kirtley Mather, "Scopes, John T./Scopes Trial."

24 Frank Thone to Donald Glassman, November 14, 1925, Science Service Records (RU7091), Box 83, Folder 1.

25 Frank Thone to Donald Glassman, January 26, 1926, Science Service Records (RU7091), Box 83, Folder 1.

26 "Weekly Bulletin, January 3, 1926, All Souls Unitarian Church," Science Service Records (RU7091), Box 443, Folder 9.

27 Typescript of Watson Davis address, January 3, 1926, Science Service Records (RU7091), Box 443, Folder 9.

28 "Evolution Foes to Conduct Fight by Aid of Radio," *Atlanta Constitution*, January 24, 1926. See also typescript of E. E. Slosson address on "Educational and Religious Freedom," May 24, 1926, Science Service Records (RU7091), Box 48, Folder 3.

29 Membership numbers cited in Virginia Gray, "Anti-Evolution Sentiment and Behavior: The Case of Arkansas," *Journal of American History* 57 (September 1970): 352–366.

30 Frank Thone correspondence with J. Willett Hill, 1926–1927, Science Service Records (RU7091), Box 86, Folder 2. See also Frank Thone correspondence with Arthur Garfield Hays, 1927, Science Service Records (RU7091), Box 85, Folder 9.

31 For example, see "Anti-Darwin Law Wins," *Los Angeles Times*, January 16, 1927; and "Scopes Is Disappointed," *New York Times*, January 16, 1925.

32 J. T. Scopes to Frank Thone, [no date on letter other than "Sunday" but it was received in the Science Service office on February 8, 1927], Science Service Records (RU7091), Box 90, Folder 5. Scopes was referring to the local chapter house of the Gamma Alpha Graduate Scientific Fraternity. See also John T. Scopes to Kirtley F. Mather, January 24, 1927, Denison University Archives, 12P M1 Box 19, K. B. Bork Biography of Kirtley Mather, "Scopes, John T./Scopes Trial."

33 Frank Thone to Arthur Garfield Hays, February 9, 1927, Science Service Records (RU7091), Box 85, Folder 9.

34 Signed and approved expense sheet, Science Service Records (RU7091), Box 90, Folder 5.

35 J. T. Scopes to Kirtley Mather, April 9, 1927, Science Service Records (RU7091), Box 90, Folder 5.

36 Scopes later wrote with affection about the friendships of those days in Chicago. And in reviewing *D-Days at Dayton* in 1966, John Stark, who had also shared a graduate student office with Scopes, mentioned that it was "particularly gratifying to note the respect and admiration shown by all the authors for one whose friendship I think of as among life's happiest experiences." John Stark, "Scopes Trial Vividly Pictured in Essays by Several Authorities," *Jackson (Tenn.) Sun* (no date on clipping), Science Service Records (97–020), Box 1, Folder 2.

37 John T. Scopes with James Presley, *Center of the Storm: Memoirs of John T. Scopes* (New York: Holt, Rinehart and Winston, 1967).

38 Frank Thone to John T. Scopes, April 16, 1927, Science Service Records (RU7091), Box 90, Folder 5.

39 Frank Thone to John T. Scopes, April 28, 1927, and John T. Scopes to Frank Thone, May 14, 1927, Science Service Records (RU7091), Box 90, Folder 5. See also John T. Scopes to Kirtley F. Mather, June 20, [1927], Denison University Archives, 12P M1 Box 19, K. B. Bork Biography of Kirtley Mather, "Scopes, John T./Scopes Trial."

40 Frank Thone to John T. Scopes, May 17, 1927, Science Service Records (RU7091), Box 90, Folder 5.

41 According to T. H. Watkins, such weaknesses were pervasive then. See T. H. Watkins, *The Hungry Years: A Narrative History of the Great Depression in America* (New York: Henry Holt and Company, 1999), 42.

42 J. T. Scopes to Frank Thone, late May 1927, Science Service Records (RU7091), Box 90, Folder 5.

43 J. T. Scopes to Frank Thone, June 20, 1927, Science Service Records (RU7091), Box 90, Folder 5.

44 Charles McD. Puckette to Frank Thone, August 15, 1927, Science Service Records (RU7091), Box 88, Folder 6. Puckette had just erroneously stated in an article that Scopes was assistant state geologist in Illinois. Charles McD. Puckette, "In Dayton Evolution Is a Dead Issue," *New York Times Sunday Magazine*, August 14, 1927.

45 Gray, "Anti-Evolution Sentiment and Behavior: The Case of Arkansas," 352–366; R. Halliburton Jr., "The Adoption of Arkansas' Anti-Evolution Law," *Arkansas Historical Quarterly* 23 (Spring 1964): 271–283; and Cal Ledbetter Jr., "The Antievolution Law: Church and State in Arkansas," *Arkansas Historical Quarterly* 38 (Winter 1979): 299–328.

46 Forrest Bailey to Frank Thone, November 9, 1928, Science Service Records

(RU7091), Box 93, Folder 1. See also Frank Thone correspondence with Roger Baldwin, Science Service Records (RU7091), Box 101, Folder 6.

47 Frank Bird to Arthur Garfield Hays, October 4, 1929, Science Service Records (RU7091), Box 101, Folder 9. Arthur O. Lovejoy, head of the American Association of University Professors, also thought the request "too canny": "If he is a qualified man, he should have no difficulty in obtaining the fellowship or assistantship which he hopes for; if he is not qualified, no university ought to give them to him; and I don't see that his qualifications can be inferred in advance merely from the fact that he is willing to undergo martyrdom if it is guaranteed to be painless—and even profitable." A. O. Lovejoy to Frank Thone, October 16, 1929, Science Service Records (RU7091), Box 105, Folder 1.

48 Frank Thone to Arthur O. Lovejoy, October 11, 1929, Science Service Records (RU7091), Box 105, Folder 1.

CHAPTER 10. LAST AND FIRST ACTS

1 E. Haldeman-Julius, *America: The Greatest Show on Earth*, Little Blue Book no. 1291 (Girard, Kans.: Haldeman-Julius Publications, 1928), 5.

2 J. T. Scopes to Frank Thone, March 7, 1931; Frank Thone to John T. Scopes, March 25, 1931; and Frank Thone to John T. Scopes, May 4, 1931, Science Service Records (RU7091), Box 129, Folder 12.

3 See, for example, Frank Thone to Winterton C. Curtis, March 12, 1931, Science Service Records (RU7091), Box 123, Folder 4; and J. Harlen Bretz to Frank Thone, December 7, 1931, Science Service Records (RU7091), Box 122, Folder 7.

4 "Scopes of Evolution Fame Seeks Congress Seat," *Chicago Daily Tribune*, August 13, 1932; "Scopes Named in House Race," *Los Angeles Times*, August 13, 1932; "'Monkey Trial' Figure Named for Congress," *New York Times*, August 13, 1932; and "Kentucky Official Majority Is 185,858," *Washington Post*, November 29, 1932.

5 See, for example, "Genesis Retained," *New York Times*, February 24, 1935; "Won't Talk Monkeys," *Chicago Daily Tribune*, September 26, 1935; and Charles F. Potter, "Ten Years after the Monkey Show I'm Going Back to Dayton," *Liberty*, September 28, 1935, 36–38. In 1935, Potter wrote authoritatively about Scopes but did not even know where he was living or what he was doing. "'Monkey Trial' Big News 25 Years Ago," *Los Angeles Times*, July 9, 1950.

6 See Watson Davis correspondence with Ray Ginger and L. Sprague De Camp in Science Service Records (RU7091), Box 333, and Science Service Records (97-020). In 1957, Davis told De Camp he could not provide information on either Scopes or Rappleyea.

7 Jerome Laurence and Robert E. Lee, *Inherit the Wind*, rev. ed. (New York: Dramatists Play Service, 1963), 3. Watson Davis eventually saw the New York pro-

duction. See Watson Davis to Kirtley Mather, April 27, 1955, Science Service Records (RU7091), Box 322, Folder 4.

8 Movie promotional brochure, Science Service Records (97–020), Box 1, Folder 3.

9 Harold J. Salemson to Watson Davis, June 28, 1960, Science Service Records (97–020), Box 1, Folder 3. Salemson, who was coordinating the event for United Artists, was well aware that Davis was a journalist and clearly hoped that he could assist with favorable publicity to the event.

10 Watson Davis, handwritten notes, July 1955, Science Service Records (97–020), Box 1, Folder 3.

11 Ibid.

12 E. E. Slosson to Herbert Rose, August 6, 1925, Science Service Records (RU7091), Box 30, Folder 1.

SOURCES

My general description of the events surrounding the trial of John Thomas Scopes during May, June, and July 1925 summarizes information in dozens of publications, all of which are listed below but not all of which agree. Where contradictions occurred, I have looked first to statements in original correspondence and documents in the Smithsonian Institution Archives, Library of Congress, University of Tennessee Special Collections, and other archival repositories and then to the accounts published by attorneys, expert witnesses, reporters, and participants such as Scopes himself.

ARCHIVAL COLLECTIONS

Manuscript Division, Library of Congress
- American Civil Liberties Union Archives: The Roger Baldwin Years, 1917–1950 (microfilm) (MSS 84156)
- Papers of William Jennings Bryan, 1877–1940 (MSS 14217)
- Papers of Luther Burbank, 1830–1989 (MSS 14324)
- James McKeen Cattell Papers (MSS 15412)
- Papers of Raymond Clapper, 1908–1960 (MSS 15925)
- Clarence Seward Darrow Papers (MSS 17756)
- Benjamin C. and Sidonie Matsner Gruenberg Papers (MSS 24265)
- Papers of Roy Wilson Howard, 1911–1966 (MSS 26583)
- John C. Merriam Papers (MSS 32706)

Smithsonian Institution Archives
- Office of the Secretary, Records, 1890–1929 (Record Unit 45)
- Office of the Secretary, Records, 1925–1949 (Record Unit 46)
- Science Service Records (Record Unit 7091)
- Science Service Records, 1920s–1970s (Accession 90-105)
- Science Service Records, 1925–1966 (Accession 97-020)

Special Collections Library, University of Tennessee, Knoxville
- John R. Neal Collection (MS 2014)
- John R. Neal Papers, 1850–1959 (MS 1126)
- Dr. James R. Montgomery Papers, 1887–1990 (MS 1880)
- The Nellie Kenyon Scrapbooks, 1917–1967 (MS 0789)
- Sue K. Hicks Papers, 1925–1975 (MS 1018)
- Sue K. Hicks and W. E. Robinson Photographic Collection

National Academies Archives

Central Policy Files, 1919–1923

Central Policy Files, 1924–1931

BIBLIOGRAPHY

Adamson, June. "Nellie Kenyon and the Scopes 'Monkey Trial.'" *Journalism History* 2 (Autumn 1975): 88–89, 97.

Allem, Warren. "Background of the Scopes Trial at Dayton, Tennessee." MA thesis, University of Tennessee, Knoxville, 1959.

Allen, Leslie H., ed. *Bryan and Darrow at Dayton: The Record of Documents of the "Bible-Evolution Trial."* New York: Russell and Russell, 1925.

"The Anti-Evolution Trial in Tennessee." *Science* 62 (July 17, 1925): 52.

Bailey, Kenneth K. "The Enactment of Tennessee's Antievolution Law." *Journal of Southern History* 16 (November 1950): 472–490.

Burke, Peter. *Eyewitnessing: The Uses of Images as Historical Evidence*. Ithaca, N.Y.: Cornell University Press, 2001.

Campbell, T. J. *Records of Rhea: A Condensed County History*. Dayton, Tenn.: Rhea Publishing, 1940.

Caudill, Edward. *Darwinism in the Press: The Evolution of an Idea*. Hillsdale, N.J.: L. Erlbaum Associates, 1989.

Clark, Constance Areson. "Evolution for John Doe: Pictures, the Public, and the Scopes Trial Debate." *Journal of American History* 87 (March 2001): 1275–1303.

Cole, Fay-Cooper. "A Witness at the Scopes Trial." *Scientific American* 200 (January 1959): 120–130.

Conkin, Paul K. *When All the Gods Trembled: Darwinism, Scopes, and American Intellectuals*. Lanham, Md.: Rowman and Littlefield, 1998.

Curtis, Winterton C. *Fundamentalism vs. Evolution at Dayton, Tennessee*, reprinted from "The Falmouth Enterprise," July 20 to August 31, 1956. N.p.: self-published, 1956.

Darrow, Clarence. *The Famous Examination of Bryan at the Scopes Evolution Trial*. Little Blue Book no. 1424. Girard, Kans.: Haldeman-Julius Publications, 1929.

———. *The Story of My Life*. New York: Charles Scribner's Sons, 1932.

Davis, Edward B. "Science and Religious Fundamentalism in the 1920s." *American Scientist* 93 (May–June 2005): 253–260.

Davis, Watson. "The Rocks and Hills of Dayton Testify for Evolution." *Science News-Letter* 6, June 25, 1925.

———. "Testimony for Evolution." *Science News-Letter* 7, July 30, 1960.

De Camp, L. Sprague. *The Great Monkey Trial*. Garden City, N.Y.: Doubleday, 1968.

Eigelsbach, William B., and Jamie Sue Linder. "'If Not the People Who?': Prosecution Correspondence Preparatory to the Scopes Trial." *Journal of East Tennessee History*, no. 70 (1998): 109–145.

Enz, Gary R. "Religion in Kansas." *Kansas History* 28 (Summer 2005): 120–145.
"Evolution, the Court, and the Church." *Scientific Monthly* 21 (December 1925): 669–670.
Gamson, Joshua. *Claims to Fame: Celebrity in Contemporary America*. Berkeley: University of California Press, 1994.
Ginger, Ray. *Six Days or Forever? Tennessee vs. John Thomas Scopes*. New York: Oxford University Press, 1958.
Gray, Virginia. "Anti-Evolution Sentiment and Behavior: The Case of Arkansas." *Journal of American History* 57 (September 1970): 352–366.
Grebstein, Norman, ed. *Monkey Trial: The State of Tennessee vs. John Thomas Scopes*. Boston: Houghton Mifflin, 1960.
Haldeman-Julius, E. *America: The Greatest Show on Earth*. Little Blue Book no. 1291. Girard, Kans.: Haldeman-Julius Publications, 1928.
———. *Lessons Life Has Taught Me: Glimpses at the Fascinating Circus of Clowns and Philosophers*. Girard, Kans.: Haldeman-Julius Publications, 1928.
Haldeman-Julius, Marcet. "Impressions of the Scopes Trial." *Haldeman-Julius Monthly* 2 (September 1925): 323–347.
———."Sinclair Lewis and a Liberal Preacher." In Henry W. Thurston (with a Symposium on Churches by Various Writers), *Why I Did Not Enter the Methodist Ministry*. Little Blue Book no. 1217, 29–58. Girard, Kans.: Haldeman-Julius Publications, 1927.
Halliburton, R., Jr. "The Adoption of Arkansas' Anti-Evolution Law." *Arkansas Historical Quarterly* 23 (Spring 1964): 271–283.
Harrison, S. L. "The Scopes 'Monkey Trial' Revisited: Mencken and the Art of Edmund Duffy." *Journal of American Culture* 17 (1994): 55–63.
Hays, Arthur Garfield. *Let Freedom Ring*. New York: Boni and Liveright, 1937.
Holt, John B. "Holiness Religion: Cultural Shock and Social Reorganization." *American Sociological Review* 5 (October 1940): 740–747.
Israel, Charles A. *Before Scopes: Evangelicalism, Education, and Evolution in Tennessee, 1870–1925*. Athens: University of Georgia Press, 2004.
Kazin, Michael. *A Godly Hero: The Life of William Jennings Bryan*. New York: Alfred A. Knopf, 2006.
Keith, Jeannette. *Country People in the New South: Tennessee's Upper Cumberland*. Chapel Hill: University of North Carolina Press, 1995.
Kent, Frank R. "On the Dayton Firing Line." *New Republic*, July 29, 1925, 259–260.
Krutch, Joseph Wood. *More Lives than One*. New York: William Sloan Associates, 1962.
LaFollette, Marcel C., ed. *Creationism, Science, and the Law: The Arkansas Case*. Cambridge, Mass.: MIT Press, 1983.
Larson, Edward. *Summer for the Gods: The Scopes Trial and America's Continuing Debate over Science and Religion*. Cambridge, Mass.: Harvard University Press, 1997.

Laurence, Jerome, and Robert E. Lee. *Inherit the Wind*. Rev. ed. New York: Dramatists Play Service, 1963.

Ledbetter, Cal, Jr. "The Antievolution Law: Church and State in Arkansas." *Arkansas Historical Quarterly* 38 (Winter 1979): 299–328.

Lienesch, Michael. *In the Beginning: The Scopes Trial and the Making of the Antievolution Movement*. Chapel Hill: University of North Carolina Press, 2007.

Mencken, H. L. *A Religious Orgy in Tennessee: A Reporter's Account of the Scopes Monkey Trial*. Hoboken, N.J.: Melville House Publishing, 2006.

Moran, Jeffrey P. *The Scopes Trial: A Brief History with Documents*. New York: Palgrave, 2002.

———. "The Scopes Trial and Southern Fundamentalism in Black and White: Race, Region, and Religion." *Journal of Southern History* 70 (February 2004): 95–120.

Morgan, W. C. "The Dayton Coal and Iron Company Limited." In Bettye J. Broyles, ed., *History of Rhea County, Tennessee*, 103–110. Dayton, Tenn.: Rhea County Historical and Genealogical Society, 1991.

Nash, Roderick. *The Nervous Generation: American Thought, 1917–1930*. New York: Rand McNally, 1970.

Newman, Horatio Hackett. *Evolution, Genetics, and Eugenics*. 3rd ed. Chicago: University of Chicago Press, 1938.

Numbers, Ronald L. "Creation, Evolution, and Holy Ghost Religion: Holiness and Pentecostal Responses to Darwinism." *Religion and American Culture* 2 (Summer 1992): 127–158.

———. *The Creationists: The Evolution of Scientific Creationism*. New York: Alfred A. Knopf, 1992.

Osborn, Henry Fairfield. "Evolution and Education in the Tennessee Trial." *Science* 62 (July 17, 1925): 43–45.

———. *Evolution and Religion in Education: Polemics of the Fundamentalist Controversy of 1922 to 1926*. New York: Charles Scribner's Sons, 1926.

Owen, Russell D. "From 'Monkey Trial' to Atomic Age." *New York Times Magazine*, July 21, 1946, 17, 37–38.

———. "The Significance of the Scopes Trial—I. Issues and Personalities." *Current History*, September 1925, 875–883.

Pauly, Philip J. *Biologists and the Promise of American Life: From Meriwether Lewis to Alfred Kinsey*. Princeton, N.J.: Princeton University Press, 2000.

Ponce de Leon, Charles A. *Self-Exposure: Human-Interest Journalism and the Emergence of Celebrity in America, 1890–1940*. Chapel Hill: University of North Carolina Press, 2002.

Potter, Charles Francis. *The Preacher and I: An Autobiography*. New York: Crown Publishers, 1951.

Scates, S. E. *A School History of Tennessee*. Yonkers-on-Hudson, N.Y.: World Book, 1925.

Schaffer, David N. *Why John Thomas Scopes Will Win His Case, and Why the Anti-Evolution Law of Tennessee Will Be Repealed*. Chicago: privately printed, 1925.

Schickel, Richard. *Intimate Strangers: The Culture of Celebrity*. Garden City, N.Y.: Doubleday, 1985.

Scopes, Jack. "Media Hysteria: Off-the-Wall Coverage of the Famous Monkey Trial," *Media History Digest* 8 (Spring-Summer 1988): 26–31.

Scopes, John T. "The Trial That Rocked the Nation." *Reader's Digest* 78, March 1961, 136–144.

Scopes, John T., with James Presley. *Center of the Storm: Memoirs of John T. Scopes*. New York: Holt, Rinehart and Winston, 1967.

The Scopes Trial: A Photographic History, with introduction by Edward Caudill and captions by Edward Larson. Knoxville: University of Tennessee Press, 2000.

Tompkins, Jerry R., ed. *D-Days at Dayton: Reflections on the Scopes Trial*. Baton Rouge: Louisiana State University Press, 1965.

Trachtenberg, Alan. *Reading American Photographs: Images as History, Mathew Brady to Walker Evans*. New York: Hill and Wang, 1989.

Wacker, Grant. *Heaven Below: Early Pentecostals and American Culture*. Cambridge, Mass.: Harvard University Press, 2001.

Warburton, T. Rennie. "Holiness Religion: An Anomaly of Sectarian Typologies." *Journal for the Scientific Study of Religion* 8 (Spring 1969): 130–139.

Wesolowski, James Walter. "Before Canon 35: WGN Broadcasts the Monkey Trial." *Journalism History* 2 (Autumn 1975): 76–79, 86–87.

WGN: A Pictorial History. Chicago: WGN, 1961.

Why Dayton—Of All Places? Dayton, Tenn.: privately printed, 1925.

Williams, Michael. "At Dayton, Tennessee." *Commonweal*, July 22, 1925, 262.

Wooten, Dudley G. *The Scopes Case*, reprinted from *Notre Dame Lawyer*, November 1925. South Bend, Ill.: Notre Dame University, 1926.

The World's Most Famous Court Trial: Tennessee Evolution Case. 3rd ed. Cincinnati, Ohio: National Book Co., 1925.

NOTES ON THE PHOTOGRAPHS

After discovery in June 2005 of a large group of original nitrate negatives within the partially processed Science Service Records at the Smithsonian Institution Archives, the archives used a grant from the Smithsonian Women's Committee to conserve a group of images related to the Scopes trial, twenty-six of which are included in this book. Nitrate film requires special handling, and so the preservation process involved creation of new safety film internegatives and a set of master prints available for digitization. The Watson Davis photographs in this book are reproduced directly from these new negatives and prints.

Figure P.1. George Washington Rappleyea and John Thomas Scopes, Dayton, Tennessee, June 1925. New print from internegative created from original nitrate negative. Smithsonian Institution Archives, Science Service Records, (RU7091), print no. 7091Davis33.

Figure 1.1. William Jennings Bryan and John Washington Butler, Dayton, Tennessee, July 1925. Scan made from print in University of Tennessee at Knoxville, Special Collections Library, Sue K. Hicks and W. E. Robinson Photographic Collection.

Figure 1.2. Robinson's Drugstore, Main Street, Dayton, Tennessee, June 1925. New print from internegative created from original nitrate negative. Smithsonian Institution Archives, Science Service Records (RU7091), print no. 7091Davis12.

Figure 1.3. Rhea County High School, Dayton, Tennessee, June 1925. New print from internegative created from original nitrate negative. Smithsonian Institution Archives, Science Service Records (RU7091), print no. CAW3.

Figure 1.4. Rhea County Courthouse, Market Street, Dayton, Tennessee, June 1925. New print from internegative created from original nitrate negative. Smithsonian Institution Archives, Science Service Records (RU7091), print no. 7091Davis39.

Figure 2.1. Edwin Emery Slosson, director, Science Service, ca. 1925. Scan made from print in Smithsonian Institution Archives, Science Service Records (Accession 90-105), Box 21, "Slosson" folder.

Figure 2.2. Watson Davis, managing editor, Science Service, ca. 1924. Scan made from print in Smithsonian Institution Archives, Science Service Records (RU7091), Box 408, Folder 1.

Figure 2.3. Frank Thone, senior biology editor, Science Service, ca. 1925. Scan made from print in Smithsonian Institution Archives, Science Service Records (Accession 90-105), Box 23, Folder "Thone."

Figure 3.1. Main Street, Dayton, Tennessee, June 1925. New print from internegative created from original nitrate negative. Smithsonian Institution Archives, Science Service Records (RU7091), print no. 7091Davis16.

Figure 3.2. John Thomas Scopes, Dayton, Tennessee, June 1925. New print from internegative created from original nitrate negative. Smithsonian Institution Archives, Science Service Records (RU7091), print no. 7091Davis35.

Figure 3.3. John Thomas Scopes and George Washington Rappleyea, Dayton, Tennessee, June 1925. New print from internegative created from original nitrate negative. Smithsonian Institution Archives, Science Service Records (RU7091), print no. 7091Davis37.

Figure 3.4. Interview notes, Watson Davis, Dayton, Tennessee, June 1925. Scan made from original notes, Smithsonian Institution Archives, Science Service Records (RU7091), Box 44, Folder 4, page headed "F. E. Robinson."

Figure 3.5. Interview notes, Watson Davis, Dayton, Tennessee, June 1925. Scan made from original notes, Smithsonian Institution Archives, Science Service Records (RU7091), Box 44, Folder 4, page headed "Top Mt. (Walden's Ridge)."

Figure 3.6. Cumberland Coal & Iron Company buildings, Dayton, Tennessee, June 1925. New print from internegative created from original nitrate negative. Smithsonian Institution Archives, Science Service Records (RU7091), print no. CAW4.

Figure 3.7. Rural countryside near Dayton, Tennessee, June 1925. New print from internegative created from original nitrate negative. Smithsonian Institution Archives, Science Service Records (RU7091), print no. CAW8.

Figure 3.8. View of Dayton, Tennessee, from tower, June 1925. New print from internegative created from original nitrate negative. Smithsonian Institution Archives, Science Service Records (RU7091), print no. 7091Davis38.

Figure 4.1. Ova Corvin Rappleyea on the steps of Defense Mansion, Dayton, Tennessee, July 1925. New print from internegative created from original nitrate negative. Smithsonian Institution Archives, Science Service Records (RU7091), print no. 7091Davis43.

Figure 4.2. Publisher E. Haldeman-Julius standing in front of Defense Mansion, Dayton, Tennessee, July 1925. New print from internegative created from original nitrate negative. Smithsonian Institution Archives, Science Service Records (RU7091), print no. 7091Davis3.

Figure 4.3. E. and Marcet Haldeman-Julius with Clarence and Ruby Darrow, Girard, Kansas, 1926. Scan made from print in Pittsburg State University, Leonard H. Axe Library, Special Collections, E. Haldeman-Julius Collection.

Figure 4.4. John Thomas Scopes and his father, Thomas Scopes, July 1925. Scan made from original glass plate negative in the Smithsonian Institution, National Museum of American History, Photographic History Collections, Underwood and Underwood Collection, photograph no. C.67.88.4922.

Figure 4.5. George Washington Rappleyea, Dayton, Tennessee, July 1925. Scan made from print in Smithsonian Institution Archives, Science Service Records (Accession 90-105), Box 18, Folder 5.

Figure 4.6. Wallace Haggard, Gordon MacKenzie, John Randolph Neal, and George Washington Rappleyea, Dayton, Tennessee, July 1925. Scan made from print in University of Tennessee at Knoxville, Special Collections Library, Sue K. Hicks and W. E. Robinson Photographic Collection.

Figure 5.1. Howard Gale Byrd, on grounds of Defense Mansion, Dayton, Tennessee, July 1925. New print from internegative created from original nitrate negative. Smithsonian Institution Archives, Science Service Records (RU7091), print no. 7091Davis31.

Figure 5.2. Group on steps of Dayton Methodist Episcopal Church (North), Dayton, Tennessee, July 1925. New print from internegative created from original nitrate negative. Smithsonian Institution Archives, Science Service Records (RU7091), print no. 7091Davis6.

Figure 5.3. George Washington Rappleyea, Howard Gale Byrd, and Charles Francis Potter, Dayton, Tennessee, July 1925. New print from internegative created from original nitrate negative. Smithsonian Institution Archives, Science Service Records (RU7091), print no. 7091Davis7.

Figure 5.4. Howard Gale Byrd and Charles Francis Potter, with Byrd's children, Dayton, Tennessee, July 1925. New print from internegative created from original nitrate negative. Smithsonian Institution Archives, Science Service Records (RU7091), print no. 7091Davis42.

Figure 5.5. Temporary privies built behind Rhea County Courthouse, July 1925. Scan made from print in Smithsonian Institution Archives, Science Service Records (RU7091), Box 404, Folder 10.

Figure 6.1. Maynard M. Metcalf, ca. 1925. Scan made from print in Smithsonian Institution Archives, Science Service Records (Accession 90-105), Box 14, Folder 42 (Metcalf).

Figure 6.2. Fay-Cooper Cole, Dayton, Tennessee, July 1925. Scan made from print in Smithsonian Institution Archives, Science Service Records (Accession 90-105), Box 5, Folder "Cohn–Colgate," photo no. 17386A.

Figure 6.3. Jacob G. Lipman, Dayton, Tennessee, July 1925. Scan made from print in Smithsonian Institution Archives, Science Service Records (Accession 90-105), Box 26, Folder "Lio–Lis," photo no. 17405.

Figure 6.4. Wilbur A. Nelson, Dayton, Tennessee, July 1925. Scan made from print in Smithsonian Institution Archives, Science Service Records (Accession 90-105), Box 16, Folder "Nelson, Norman," photo no. 17417.

Figure 6.5. Winterton C. Curtis, Dayton, Tennessee, July 1925. Scan made from print in Smithsonian Institution Archives, Science Service Records (Accession 90-105), Box 4, Folder "Winterton C. Curtis," photo no. 17389.

Figure 6.6. William A. Kepner, Dayton, Tennessee, July 1925. Scan made from print in Smithsonian Institution Archives, Science Service Records (Accession 90-105), Box 11, Folder "Portraits, Keo-Kep," photo no. 17404.

Figure 6.7. Scientists gathered at Defense Mansion, Dayton, Tennessee, July 1925. Smithsonian Institution Archives. New print from internegative created from original nitrate negative. Smithsonian Institution Archives, Science Service Records (RU7091), CAW 10, NV Box 360, Folder 1, Sheet 1, negative 11.

Figure 6.8. Scientists, theologians, journalists, and defense lawyers, sitting on steps of Defense Mansion, Dayton, Tennessee, July 1925. Scan made from print in Smithsonian Institution Archives, Science Service Records (Accession 97-020), Box 1, Folder 2.

Figure 6.9. Defense attorneys, journalists, and expert witnesses assembled for the Scopes trial, Dayton, Tennessee, July 1925. Scan made from print in University of Tennessee at Knoxville, Special Collections Library, Sue K. Hicks and W. E. Robinson Photographic Collection.

Figure 6.10. Letter to Michael Pupin from the "scientists at Dayton," July 17, 1925. Photograph made from original document in Library of Congress, Clarence Seward Darrow Papers, Container 5, Folder 1.

Figure 7.1. Worshippers assembled for a baptism in a stream near Dayton, Tennessee, July 19, 1925. New print from internegative created from original nitrate negative. Smithsonian Institution Archives, Science Service Records (RU7091), print no. 7091Davis50.

Figure 7.2. Worshippers assembled for a baptism in a stream near Dayton, Tennessee, July 19, 1925. New print from internegative created from original nitrate negative. Smithsonian Institution Archives, Science Service Records (RU7091), print no. 7091Davis51.

Figure 7.3. Worshippers assembled for a baptism in a stream near Dayton, Tennessee, July 19, 1925. New print from internegative created from original nitrate negative. Smithsonian Institution Archives, Science Service Records (RU7091), print no. 7091Davis47.

Figure 7.4. Worshippers observing a baptism near Dayton, Tennessee, July 19, 1925. New print from internegative created from original nitrate negative. Smithsonian Institution Archives, Science Service Records (RU7091), print no. 7091Davis45.

Figure 7.5. Worshippers observing a baptism near Dayton, Tennessee, July 19, 1925. New print from internegative created from original nitrate negative. Smithsonian Institution Archives, Science Service Records (RU7091), print no. 7091Davis48.

Figure 7.6. Worshippers observing a baptism near Dayton, Tennessee, July 19, 1925. New print from internegative created from original nitrate negative. Smithsonian Institution Archives, Science Service Records (RU7091), print no. 7091Davis44.

Figure 8.1. Clarence Darrow's interrogation of William Jennings Bryan, Dayton, Tennessee, July 20, 1925. New print from internegative created from original nitrate negative. Smithsonian Institution Archives, Science Service Records (RU7091), print no. 7091Davis17.

Figure 8.2. Clarence Darrow's interrogation of William Jennings Bryan, Dayton, Tennessee, July 20, 1925. New print from internegative created from original nitrate negative. Smithsonian Institution Archives, Science Service Records (RU7091), print no. 7091Davis18.

Figure 8.3. Clarence Darrow's interrogation of William Jennings Bryan, Dayton, Tennessee, July 20, 1925. New print from internegative created from original nitrate negative. Smithsonian Institution Archives, Science Service Records (RU7091), print no. 7091Davis28.

Figure 8.4. William Jennings Bryan's coffin being loaded onto train, Dayton, Tennessee, July 1925. Scan made from print in University of Tennessee at Knoxville, Special Collections Library, Sue K. Hicks and W. E. Robinson Photographic Collection.

Figure 9.1. Outside cover of bank book for John T. Scopes Scholarship Fund. Scan of object in Smithsonian Institution Archives, Science Service Records (RU7091), Box 90, Folder 5.

Figure 9.2. Ledger pages from bank book for John T. Scopes Scholarship Fund. Scan of object in Smithsonian Institution Archives, Science Service Records (RU7091), Box 90, Folder 5.

Figure 9.3. Letter from John Thomas Scopes to Frank Thone, June 20, 1927, page 1. Scan made from original document in Smithsonian Institution Archives, Science Service Records (RU7091), Box 90, Folder 5.

Figure 9.4. Letter from John Thomas Scopes to Frank Thone, June 20, 1927, page 2. Scan made from original document in Smithsonian Institution Archives, Science Service Records (RU7091), Box 90, Folder 5.

ACKNOWLEDGMENTS

Historians interpret, ask questions, reconstruct events, piece together stories. We cannot do these jobs, however, without history's raw materials—the images, written records, and objects acquired and preserved by archives and archivists, special collections and librarians, museums and curators. Many of the images, letters, and documents used in this book were almost lost forever. Thanks to the foresight and fortuitous intervention of the late Audrey Davis (a curator and historian at the Smithsonian Institution and daughter-in-law to Watson Davis) and to the actions of other Smithsonian curators and archivists, the early editorial records of Science Service (including the box in which the Scopes trial negatives were found) were saved from destruction during the late 1960s (and other sets in subsequent years). The Smithsonian Institution Archives (SIA) eventually accessioned these materials, but they arrived at a time when expanding administrative records and decreasing budgets meant less staff time available for processing the more complex and eclectic collections. Hundreds upon hundreds of boxes remained uncataloged for decades, and as a result, the Science Service records were little used, their potential treasures left unmined.

Ellen Alers, Tammy Peters, Shawn Johnstone, Jim Steed, Sarah Stauderman, and the rest of the wonderful SIA staff, past and present, have been an integral part of the research that produced this book. I cannot adequately express my thanks for all their support, suggestions, and good humor (especially the good humor) nor sufficiently describe the extent of my admiration for their work. Being an archivist today involves not just intelligence and creativity but heavy lifting and dedication to the cause of finding the truth, even if that involves an occasional daredevil scramble to the highest shelves in a warehouse.

Unfortunately, the institutions so necessary for preserving our common cultural resources are increasingly in peril. Without adequate funding, without ingenious and dedicated staff, without funding for restoration, without sufficient space for storage and conservation activities, precious collections can deteriorate or, worse, never be preserved at all. The survival and restoration of the Scopes trial photographs should serve as both a lesson about the importance of preservation and a call to increase support for archives and libraries everywhere.

In addition to the staff at the Smithsonian, I thank the archivists at the Special Collections Library, University of Tennessee at Knoxville, especially Bill Eigelsbach, Nick Wyman, and Elizabeth Dunham, for their help at the early stages of this research. I also thank the resourceful archivists at the Library of Congress Manuscript Division, Janice Goldblum at the National Academies Archives, and Heather Lyle, University Archivist and Special Collections Librarian at Denison University.

I am indebted to Shannon Perich and David Haberstich at the Smithsonian's National Museum of American History for help not only in interpreting the Davis negatives but also in arranging a new print of the John and Thomas Scopes photograph. Thank you to Randy Roberts at Special Collections, Pittsburg State University, for providing the photograph of E. Haldeman-Julius reproduced in this book and to Bonnie Coles at the Library of Congress for help in obtaining the reproduction of the letter to Michael Pupin.

Thank you to Jeffrey P. Moran for valuable comments and encouragement on the draft and to Fred Woodward of the University Press of Kansas for (more) good humor and overall support for this project.

A number of friends have shared in the excitement of my research, nourishing me with good will, jokes, and the conviviality that keeps authors afloat. Thank you, especially, to Chris Cherniak, Jonathan Cobb, Mary Cole, Jonathan Coopersmith, Patricia Garfinkel, David R. Gessner, Bart Hacker, Lisa Helperin, Don Reisman, Irene Schubert, Mickey Schubert, Margaret Vining, and Terry Young.

From the moment I first uncovered the Scopes negatives, my husband, Jeffrey Stine, has shared in the joy, excitement, and fascination. He did not hesitate to encourage me to move forward with this project and has offered enthusiastic support at every step. Jeffrey, you always believed that this book would be possible—but it would not have been possible without you.

—Marcel C. LaFollette

INDEX

Page numbers in italics refer to illustrations.